MIND AND COSMOS

BY THE SAME AUTHOR

The Possibility of Altruism
Mortal Questions
The View from Nowhere
What Does It All Mean?
Equality and Partiality
Other Minds
The Last Word
The Myth of Ownership (with Liam Murphy)
Concealment and Exposure
Secular Philosophy and the Religious Temperament

MIND AND COSMOS

Why the Materialist Neo-Darwinian Conception of Nature Is Almost Certainly False

Thomas Nagel

OXFORD
UNIVERSITY PRESS

Oxford University Press is a department of the University of Oxford.
It furthers the University's objective of excellence in research,
scholarship, and education by publishing worldwide.

Oxford New York
Auckland Cape Town Dar es Salaam Hong Kong Karachi
Kuala Lumpur Madrid Melbourne Mexico City Nairobi
New Delhi Shanghai Taipei Toronto

With offices in
Argentina Austria Brazil Chile Czech Republic France Greece
Guatemala Hungary Italy Japan Poland Portugal Singapore
South Korea Switzerland Thailand Turkey Ukraine Vietnam

Published in the United States of America by Oxford University Press
198 Madison Avenue, New York, NY 10016

Library of Congress Cataloging-in-Publication Data
Nagel, Thomas, 1937–
Mind and cosmos : why the materialist neo-Darwinian conception
of nature is almost certainly false / Thomas Nagel.
p. cm.
ISBN 978-0-19-991975-8 (alk. paper)
1. Cosmology 2. Cosmogony. 3. Beginning. 4. Creation. 5. Science—Philosophy.
6. Darwin, Charles, 1809–1882. I. Title.
B D511.N34 2012
113—dc23 2011051647

7 9 8 6

Printed in the United States of America
on acid-free paper

To Anne

CONTENTS

PREFACE

The two people who, in very different ways, have had the most influence on the thoughts expressed in this book are Sharon Street and Roger White. I was also much instructed and stimulated by the discussions of a research group on science and religion organized at the New York Institute of Philosophy with the support of a Mellon Foundation Distinguished Achievement Award, and I am grateful to the Mellon Foundation for making it possible. The group, which met from 2006 through 2009, brought together faculty and graduate students of the philosophy department at New York University with regular and occasional participants from other universities and other fields. Street and White were members of that group, and I would also like to thank in particular Paul Boghossian, Laura Franklin-Hall, Philip Kitcher, Matthew Kotzen, H. Allen Orr, Alvin Plantinga, Elliott Sober, and Michael Strevens. Sober also read the manuscript of the book for Oxford University Press, and offered useful suggestions. I presented some of the material at the Colloquium in Legal, Political, and Social Philosophy that Ronald Dworkin and I have conducted for many years, and I am grateful to him and to the other participants for their help. In view of the unorthodoxy of the result, I hope these thanks will not give offense.

During the writing of the book I received research support from the Filomen D'Agostino and Max E. Greenberg Faculty Research Fund of New York University School of Law.

New York, October, 2011

MIND AND COSMOS

Chapter 1

Introduction

The aim of this book is to argue that the mind-body problem is not just a local problem, having to do with the relation between mind, brain, and behavior in living animal organisms, but that it invades our understanding of the entire cosmos and its history. The physical sciences and evolutionary biology cannot be kept insulated from it, and I believe a true appreciation of the difficulty of the problem must eventually change our conception of the place of the physical sciences in describing the natural order.

One of the legitimate tasks of philosophy is to investigate the limits of even the best developed and most successful forms of contemporary scientific knowledge. It may be frustrating to acknowledge, but we are simply at the point in the history of human thought at which we find ourselves, and our successors will make discoveries and develop forms of understanding of which we have not dreamt. Humans are addicted to the hope for a final reckoning, but intellectual humility requires that we resist the temptation to assume that tools of the kind we now have are in principle sufficient to understand the universe as a whole. Pointing out their limits is a philosophical task, whoever engages in it, rather than part of the internal pursuit of science—though we can hope that if the limits are recognized, that may eventually lead to the discovery of new forms of scientific understanding. Scientists are well aware of how much they don't know, but this is a different kind of problem—not just of acknowledging the limits of what is actually understood but of trying to recognize what can and cannot in principle be understood by certain existing methods.

My target is a comprehensive, speculative world picture that is reached by extrapolation from some of the discoveries of biology, chemistry, and physics—a particular naturalistic *Weltanschauung* that postulates a hierarchical relation among the subjects of those sciences, and the completeness in principle of an explanation of everything in the universe through their unification. Such a world view is not a necessary condition of the practice of any of those sciences, and its acceptance or nonacceptance would have no effect on most scientific research. For all I know, most practicing scientists may have no opinion about the overarching cosmological questions to which this materialist reductionism provides an answer. Their detailed research and substantive findings do not in general depend on or imply either that or any other answer to such questions. But among the scientists and philosophers who do express views about the natural order as a whole, reductive materialism is widely assumed to be the only serious possibility.[1]

The starting point for the argument is the failure of psychophysical reductionism, a position in the philosophy of mind that is largely motivated by the hope of showing how the physical sciences could in principle provide a theory of everything. If that hope is unrealizable, the question arises whether any other more or less unified understanding could take in the entire cosmos as we know it. Among the traditional candidates for comprehensive understanding of the relation of mind to the physical world, I believe the weight of

1. For a clear statement, see Steven Weinberg, *Dreams of a Final Theory* (New York: Pantheon Books, 1992), chapter 3.

evidence favors some form of neutral monism over the traditional alternatives of materialism, idealism, and dualism. What I would like to do is to explore the possibilities that are compatible with what we know—in particular what we know about how mind and everything connected with it depends on the appearance and development of living organisms, as a result of the universe's physical, chemical, and then biological evolution. I will contend that these processes must be reconceived in light of what they have produced, if psychophysical reductionism is false.

The argument from the failure of psychophysical reductionism is a philosophical one, but I believe there are independent empirical reasons to be skeptical about the truth of reductionism in biology. Physico-chemical reductionism in biology is the orthodox view, and any resistance to it is regarded as not only scientifically but politically incorrect. But for a long time I have found the materialist account of how we and our fellow organisms came to exist hard to believe, including the standard version of how the evolutionary process works. The more details we learn about the chemical basis of life and the intricacy of the genetic code, the more unbelievable the standard historical account becomes.[2] This is just the opinion of a layman who reads widely in the literature that explains contemporary science to the nonspecialist. Perhaps that literature presents the situation with a simplicity and confidence that does not reflect the most sophisticated scientific thought in these areas. But it seems to me that, as it is usually presented, the current orthodoxy about the cosmic order is the product of governing assumptions that are unsupported, and that it flies in the face of common sense.

2. See Richard Dawkins, *The Blind Watchmaker: Why the Evidence of Evolution Reveals a Universe without Design* (New York: Norton, 1986), for a canonical exposition, which seems to convince practically everyone.

I would like to defend the untutored reaction of incredulity to the reductionist neo-Darwinian account of the origin and evolution of life.[3] It is prima facie highly implausible that life as we know it is the result of a sequence of physical accidents together with the mechanism of natural selection. We are expected to abandon this naïve response, not in favor of a fully worked out physical/chemical explanation but in favor of an alternative that is really a schema for explanation, supported by some examples. What is lacking, to my knowledge, is a credible argument that the story has a nonnegligible probability of being true. There are two questions. First, given what is known about the chemical basis of biology and genetics, what is the likelihood that self-reproducing life forms should have come into existence spontaneously on the early earth, solely through the operation of the laws of physics and chemistry? The second question is about the sources of variation in the evolutionary process that was set in motion once life began: In the available geological time since the first life forms appeared on earth, what is the likelihood that, as a result of physical accident, a sequence of viable genetic mutations should have occurred that was sufficient to permit natural selection to produce the organisms that actually exist?

There is much more uncertainty in the scientific community about the first question than about the second. Many people think it will be very difficult to come up with a reductionist explanation of the origin of life, but most people have no doubt that accidental genetic variation is enough to support the actual history of evolution by natural selection, once reproducing organisms have come into existence. However, since the questions concern highly specific events over a

3. For an illuminating account of Darwin's own views about the most basic forms of explanation, see Elliott Sober, "Darwin's Discussions of God," in *Did Darwin Write the "Origin" Backwards?: Philosophical Essays on Darwin's Theory* (Amherst, NY: Prometheus Books, 2011), 121–28.

long historical period in the distant past, the available evidence is very indirect, and general assumptions have to play an important part. My skepticism is not based on religious belief, or on a belief in any definite alternative. It is just a belief that the available scientific evidence, in spite of the consensus of scientific opinion, does not in this matter rationally require us to subordinate the incredulity of common sense. That is especially true with regard to the origin of life.

The world is an astonishing place, and the idea that we have in our possession the basic tools needed to understand it is no more credible now than it was in Aristotle's day. That it has produced you, and me, and the rest of us is the most astonishing thing about it. If contemporary research in molecular biology leaves open the possibility of legitimate doubts about a fully mechanistic account of the origin and evolution of life, dependent only on the laws of chemistry and physics, this can combine with the failure of psychophysical reductionism to suggest that principles of a different kind are also at work in the history of nature, principles of the growth of order that are in their logical form teleological rather than mechanistic. I realize that such doubts will strike many people as outrageous, but that is because almost everyone in our secular culture has been browbeaten into regarding the reductive research program as sacrosanct, on the ground that anything else would not be science.

My project has the familiar form of trying to meet a set of conditions that seem jointly impossible. In addition to antireductionism, two further constraints are important: first, an assumption that certain things are so remarkable that they have to be explained as non-accidental if we are to pretend to a real understanding of the world; second, the ideal of discovering a single natural order that unifies everything on the basis of a set of common elements and principles—an ideal toward which the inevitably very incomplete forms of our actual understanding should nevertheless aspire. Cartesian dualism rejects this second aspiration, and the reductive programs of both materialism and idealism are failed attempts to realize it. The unifying conception is also incompatible with the kind of theism that explains certain features of the natural world by divine intervention, which is not part of the natural order.

The great advances in the physical and biological sciences were made possible by excluding the mind from the physical world. This has permitted a quantitative understanding of that world, expressed in timeless, mathematically formulated physical laws. But at some point it will be necessary to make a new start on a more comprehensive understanding that includes the mind. It seems inevitable that such an understanding will have a historical dimension as well as a timeless one. The idea that historical understanding is part of science has become familiar through the transformation of biology by evolutionary theory. But more recently, with the acceptance of the big bang, cosmology has also become a historical science. Mind, as a development of life, must be included as the most recent stage of this long cosmological history, and its appearance, I believe, casts its shadow back over the entire process and the constituents and principles on which the process depends.

The question is whether we can integrate this perspective with that of the physical sciences as they have been developed for a mindless universe. The understanding of mind cannot be contained within the personal point of view, since mind is the product of a partly physical process; but by the same token, the separateness of physical science, and its claim to completeness, has to end in the long run. And that poses the question: To what extent will the reductive form that is so central to contemporary physical science survive this transformation? If physics and chemistry cannot fully account for life and consciousness, how will their immense body of truth be combined with other elements in an expanded conception of the natural order that can accommodate those things?

As I have said, doubts about the reductionist account of life go against the dominant scientific consensus, but that consensus faces problems of probability that I believe are not taken seriously enough, both with respect to the evolution of life forms through accidental mutation and natural selection and with respect to the formation from dead matter of physical systems capable of such evolution. The more we learn about the intricacy of the genetic code and its control of the chemical processes of life, the harder those problems seem.

Again: with regard to evolution, the process of natural selection cannot account for the actual history without an adequate supply of viable mutations, and I believe it remains an open question whether this could have been provided in geological time merely as a result of chemical accident, without the operation of some other factors determining and restricting the forms of genetic variation. It is no longer legitimate simply to imagine a sequence of gradually evolving phenotypes, as if their appearance through mutations in the DNA were unproblematic—as Richard Dawkins does for the evolution of the eye.[4] With regard to the origin of life, the problem is much harder, since

4. See Dawkins, *The Blind Watchmaker*, 77–86. Jerry Fodor and Massimo Piattelli-Palmarini argue in the first part of their book *What Darwin Got Wrong* (New York: Farrar, Straus & Giroux, 2010) that Darwinian evolutionary theory assigns much too much of the explanatory burden for the functional character of organisms to the external influence of natural selection, and not enough to the sources of genetic variation. This point is independent of their attack on the alleged intentionality of the idea of natural selection in the second part of the book—which seems to me, as to others, to be based on a misinterpretation.

There are also more mainstream figures who insist that the evidence calls for a more restricted account of the sources of variation in the genetic material. Marc W. Kirschner and John C. Gerhart, in *The Plausibility of Life: Resolving Darwin's Dilemma* (New Haven, CT: Yale University Press, 2005), suggest that genetic variation is biased to facilitate evolutionary change, though they do not imply that this calls for a revision of the larger reductionist conception of nature. Stuart Kauffman suggests in several books that variation is not due to chance, and that principles of spontaneous self-organization play a more important role than natural selection in evolutionary history. See *At Home in the Universe: The Search for Laws of Self-Organization and Complexity* (New York: Oxford University Press, 1995); *Investigations* (New York: Oxford University Press, 2000); *Reinventing the Sacred: A New View of Science, Reason, and Religion* (New York: Basic Books, 2008).

the option of natural selection as an explanation is not available. And the coming into existence of the genetic code—an arbitrary mapping of nucleotide sequences into amino acids, together with mechanisms that can read the code and carry out its instructions—seems particularly resistant to being revealed as probable given physical law alone.[5]

In thinking about these questions I have been stimulated by criticisms of the prevailing scientific world picture from a very different direction: the attack on Darwinism mounted in recent years from a religious perspective by the defenders of intelligent design. Even though writers like Michael Behe and Stephen Meyer are motivated at least in part by their religious beliefs, the empirical arguments they offer against the likelihood that the origin of life and its evolutionary history can be fully explained by physics and chemistry are of great interest in themselves.[6] Another skeptic, David Berlinski, has brought out these problems vividly without reference to the design inference.[7] Even if one is not drawn to the alternative of an explanation by the actions of a designer, the problems that these iconoclasts pose for the orthodox scientific consensus should be taken seriously.[8] They do not deserve the scorn with which they are commonly met. It is manifestly unfair.

5. Indeed there may be something deeply confused about the request for such an explanation—for a reason pointed out by Roger White, which I discuss in chapter 4.
6. See Michael J. Behe, *Darwin's Black Box: The Biochemical Challenge to Evolution* (New York: Simon & Schuster, 1996); Behe, *The Edge of Evolution: The Search for the Limits of Darwinism* (New York: Free Press, 2007); Stephen C. Meyer, *Signature in the Cell: DNA and the Evidence for Intelligent Design* (New York: HarperOne, 2009).
7. See David Berlinski, "On the Origins of Life," *Commentary*, February 2006, reprinted in Berlinski, *The Deniable Darwin, and Other Essays* (Seattle: Discovery Institute Press, 2009). See also Brian Goodwin, *How the Leopard Changed Its Spots: The Evolution of Complexity* (New York: Scribner's, 1994).
8. There are also criticisms of current theories from those who nevertheless expect a reductive solution; for example Robert Shapiro, *Origins: A Skeptic's Guide to the Creation of Life on Earth* (New York: Summit Books, 1986); Shapiro, "A Simpler Origin for Life," *Scientific American*, February 12, 2007. A very clear explanation of multiple aspects of current research into the origin of life and the possibility of extraterrestrial life is Steven Benner, *Life, the Universe and the Scientific Method* (Gainesville, FL: FfAME Press, 2008). Though

Those who have seriously criticized these arguments have certainly shown that there are ways to resist the design conclusion; but the general force of the negative part of the intelligent design position—skepticism about the likelihood of the orthodox reductive view, given the available evidence—does not appear to me to have been destroyed in these exchanges.[9] At least, the question should be regarded as open. To anyone interested in the basis of this judgment, I can only recommend a careful reading of some of the leading advocates on both sides of the issue—with special attention to what has been established by the critics of intelligent design. Whatever one may think about the possibility of a designer, the prevailing doctrine—that the appearance of life from dead matter and its evolution through accidental mutation and natural selection to its present forms has involved nothing but the operation of physical law—cannot be regarded as unassailable. It is an assumption governing the scientific project rather than a well-confirmed scientific hypothesis.

---

he assumes this is a task for chemistry, he does say (287), "A real potential exists that current theory will *never* solve the problem at hand, keeping open the possibility for a true revolution in the related and surrounding sciences." Of course he doesn't mean intelligent design.

A problem with the most salient current research is that the synthesis of individual components of the genetic material is so heavily controlled and guided by the experimenters that it provides little evidence that the process could have occurred without intelligent guidance. And the crucial question of how these components could have combined into an information-rich coded sequence is left unaddressed.

9. The literature is extensive. See for example Kenneth R. Miller, *Finding Darwin's God: A Scientist's Search for Common Ground between God and Evolution* (New York: Cliff Street, 1999); Philip Kitcher, *Living with Darwin: Evolution, Design, and the Future of Faith* (New York: Oxford University Press, 2007); Elliott Sober, *Evidence and Evolution: The Logic Behind the Science* (Cambridge, UK: Cambridge University Press, 2008); for a sample of both sides of the debate, see Neil A. Manson, ed., *God and Design: The Teleological Argument and Modern Science* (New York: Routledge, 2003).

I confess to an ungrounded assumption of my own, in not finding it possible to regard the design alternative as a real option. I lack the *sensus divinitatis* that enables—indeed compels—so many people to see in the world the expression of divine purpose as naturally as they see in a smiling face the expression of human feeling.[10] So my speculations about an alternative to physics as a theory of everything do not invoke a transcendent being but tend toward complications to the immanent character of the natural order. That would also be a more unifying explanation than the design hypothesis. I disagree with the defenders of intelligent design in their assumption, one which they share with their opponents, that the only naturalistic alternative is a reductionist theory based on physical laws of the type with which we are familiar. Nevertheless, I believe the defenders of intelligent design deserve our gratitude for challenging a scientific world view that owes some of the passion displayed by its adherents precisely to the fact that it is thought to liberate us from religion.

That world view is ripe for displacement, in spite of the great achievements of reductive materialism, which will presumably continue for a long time to be our main source for concrete understanding and control of the world around us. To argue, as I will, that there is a lot it can't explain is not to offer an alternative. But the recognition of those limits is a precondition of looking for alternatives, or at least of being open to their possibility. And it may mean that some directions of pursuit of the materialist form of explanation will come to be seen as dead ends. If the appearance of conscious organisms in the world is due to principles of development that are not derived from the timeless laws of physics, that may be a reason for pessimism about purely chemical explanations of the origin of life as well.

10. I am not just unreceptive but strongly averse to the idea, as I have said elsewhere.

Chapter 2

Antireductionism and the Natural Order

1

The conflict between scientific naturalism and various forms of antireductionism is a staple of recent philosophy. On one side there is the hope that everything can be accounted for at the most basic level by the physical sciences, extended to include biology.[1] On the other side there are doubts about whether the reality of such features of our world as consciousness, intentionality, meaning, purpose, thought, and value can be accommodated in a universe consisting at the most basic level only of physical facts—facts, however sophisticated, of the kind revealed by the physical sciences.

I will use the terms "materialism" or "materialist naturalism" to refer to one side of this conflict and "antireductionism" to refer to the other side, even though the terms are rather rough. The attempts to defend the materialist world picture as a potentially complete account of what there is take many forms, and not all of them involve reduction in the ordinary sense, such as the analysis of mental concepts in behavioral terms or the scientific identification of mental states with brain states. Many materialist naturalists would not describe their view

1. This program has been pursued with dedication in the writings of Daniel Dennett.

as reductionist. But to those who doubt the adequacy of such a world view, the different attempts to accommodate within it mind and related phenomena all appear as attempts to reduce the true extent of reality to a common basis that is not rich enough for the purpose. Hence the resistance can be brought together as antireductionism.

The tendency of these antireductionist doubts is usually negative. The conclusion they invite is that there are some things that the physical sciences alone cannot fully account for. Other forms of understanding may be needed, or perhaps there is more to reality than even the most fully developed physics can describe. If reduction fails in some respect, this reveals a limit to the reach of the physical sciences, which must therefore be supplemented by something else to account for the missing elements. But the situation may be more serious than that. If one doubts the reducibility of the mental to the physical, and likewise of all those other things that go with the mental, such as value and meaning, then there is some reason to doubt that a reductive materialism can apply even in biology, and therefore reason to doubt that materialism can give an adequate account even of the physical world. I want to explore the case for this breakdown, and to consider whether anything positive by way of a world view is imaginable in the wake of it.

We and other creatures with mental lives are organisms, and our mental capacities apparently depend on our physical constitution. So what explains the existence of organisms like us must also explain the existence of mind. But if the mental is not itself merely physical, it cannot be fully explained by physical science. And then, as I shall argue, it is difficult to avoid the conclusion that those aspects of our physical constitution that bring with them the mental cannot be fully explained by physical science either. If evolutionary biology is a physical theory—as it is generally taken to be—then it cannot account for the appearance of consciousness and of other phenomena that are not physically reducible. So if mind is a product of biological evolution—if organisms with mental life are not miraculous anomalies but an integral part of nature—then biology cannot be a purely physical science. The possibility opens up of a pervasive conception of the natural order very different from materialism—one that makes mind central, rather than a side effect of physical law.

It seems clear that the conclusion of antireductionist arguments against materialism cannot remain purely negative forever. Even if the dominance of materialist naturalism is nearing its end, we need some idea of what might replace it. One of the things that drive the various reductionist programs about mind, value, and meaning, in spite of their inherent implausibility, is the lack of any comprehensive alternative. It can seem that the only way to accept the arguments against reduction is by adding peculiar extra ingredients like qualia, meanings, intentions, values, reasons, beliefs, and desires to the otherwise magnificently unified mathematical order of the physical universe. But this does not answer to the desire for a general understanding of how things fit together. A genuine alternative to the reductionist program would require an account of how mind and everything that goes with it is inherent in the universe.

I am just turning a familiar argument on its head in order to challenge the premises. Materialism requires reductionism; therefore the failure of reductionism requires an alternative to materialism. My aim is not so much to argue against reductionism as to investigate the consequences of rejecting it—to present the problem rather than to propose a solution. Materialist naturalism leads to reductionist ambitions because it seems unacceptable to deny the reality of all those familiar things that are not at first glance physical. But if no plausible reduction is available, and if denying reality to the mental continues to be unacceptable, that suggests that the original premise, materialist naturalism, is false, and not just around the edges. Perhaps the natural order is not exclusively physical; or perhaps, in the worst case, there is no comprehensive natural order in which everything hangs together—only disconnected forms of understanding. But whatever may be the result, we must start out from a larger conception of what has to be understood in order to make sense of the natural world.

2

My guiding conviction is that mind is not just an afterthought or an accident or an add-on, but a basic aspect of nature. Quite apart from antireductionist arguments in the philosophy of mind, there is independent support for the step to such an enlarged conception of reality in one of the background conditions of science. Science is driven by the assumption that the world is intelligible. That is, the world in which we find ourselves, and about which experience gives us some information, can be not only described but understood. That assumption is behind every pursuit of knowledge, including pursuits that end in illusion. In the natural sciences as they have developed since the seventeenth century, the assumption of intelligibility has led to extraordinary discoveries, confirmed by prediction and experiment, of a hidden natural order that cannot be observed by human perception alone. Without the assumption of an intelligible underlying order, which long antedates the scientific revolution, those discoveries could not have been made.

What explains this order? One answer would be that nothing does: explanation comes to an end with the order itself, which the assumption of intelligibility has merely enabled us to uncover. Perhaps one level of order can be explained in terms of a still deeper level—as has happened repeatedly in the history of science. But in the end, on this view of the matter, understanding of the world will eventually reach a point where there is nothing more to be said, except "This is just how things are."

I am not disposed to see the success of science in this way. It seems to me that one cannot really understand the scientific world view unless one assumes that the intelligibility of the world, as described by the laws that science has uncovered, is itself part of the deepest explanation of why things are as they are. So when we prefer one explanation of the same data to another because it is simpler and makes fewer arbitrary assumptions, that is not just an aesthetic preference: it is because we think the explanation that gives greater understanding is more likely to be true, just for that reason.

This assumption is a form of the principle of sufficient reason—that everything about the world can at some level be understood, and that if many things, even the most universal, initially seem arbitrary, that is because there are further things we do not know, which explain why they are not arbitrary after all.

The view that rational intelligibility is at the root of the natural order makes me, in a broad sense, an idealist—not a subjective idealist, since it doesn't amount to the claim that all reality is ultimately appearance—but an objective idealist in the tradition of Plato and perhaps also of certain post-Kantians, such as Schelling and Hegel, who are usually called absolute idealists. I suspect that there must be a strain of this kind of idealism in every theoretical scientist: pure empiricism is not enough.

The intelligibility of the world is no accident. Mind, in this view, is doubly related to the natural order. Nature is such as to give rise to conscious beings with minds; and it is such as to be comprehensible to such beings. Ultimately, therefore, such beings should be comprehensible to themselves. And these are fundamental features of the universe, not byproducts of contingent developments whose true explanation is given in terms that do not make reference to mind.

3

The largest question within which all natural science is embedded is also the largest question of philosophy—namely, in what way or ways is the world intelligible? Clearly natural science is one of the most important ways of revealing intelligibility. But in spite of the great accomplishments of the natural sciences in their present form, it is important both for science itself and for philosophy to ask how much of what there is the physical sciences can render intelligible—how much of the world's intelligibility consists in its subsumability under universal, mathematically formulable laws governing the spatiotemporal order. If there are limits to the reach of science in this form, are there other forms of understanding that can render intelligible what physical science does not explain?

But first we should consider the view that there are no such limits—that physical law has the resources to explain everything, including the double relation of mind to the natural order. The intelligibility (to us) that makes science possible is one of the things that stand in need of explanation. The strategy is to try to extend the materialist world picture so that it includes such an explanation, thereby making the physical intelligibility of the world close over itself. According to this type of naturalism, the existence of minds to whom the world is scientifically intelligible is itself scientifically explicable, as a highly specific biological side effect of the physical order.

The story goes like this: There is no need for an expanded form of understanding; instead, the history of human knowledge gives us reason to believe that there is ultimately one way that the natural order is intelligible, namely, through physical law—everything that exists and everything that happens can in principle be explained by the laws that govern the physical universe. Admittedly, we can't grasp the natural order in its full manifestation because it is too complex, and we therefore need more specialized forms of understanding for practical purposes. But we can attempt to discover the universal principles governing the elements out of which everything is composed, and of which all observable spatiotemporal complexity is a manifestation. These are the mathematically stateable laws of basic physics, which describe the fundamental forces and particles or other entities and their interactions, at least till a still more fundamental level is uncovered. The most systematic possible description of a material universe extended in space and time is therefore the route to the most fundamental explanation of everything.

Physics and chemistry have pursued this aim with spectacular success. But the great step forward in the progress of the materialist conception toward the ideal of completeness was the theory of evolution, later reinforced and enriched by molecular biology and the discovery of DNA. Modern evolutionary theory offers a general picture of how the existence and development of life could be just another consequence of the equations of particle physics. Even if no one yet has a workable idea about the details, it is possible to speculate that the appearance of life was the product of chemical processes governed by the laws of physics, and that evolution after that is likewise due to chemical mutations and natural selection that are also just super-complex consequences of physical principles. Even if there is a residual problem of exactly how to account for consciousness in physical terms, the orthodox naturalistic view is that biology is in principle completely explained by physics and chemistry, and that evolutionary psychology provides a rough idea of how everything distinctive about human life can also be regarded as an extremely complicated consequence of the behavior of physical particles in accordance with certain fundamental laws. This will ultimately include an explanation of the cognitive capacities that enable us to discover those laws.

I find it puzzling that this view of things should be taken as more or less self-evident, as I believe it commonly is. Everyone acknowledges that there are vast amounts we do not know, and that enormous opportunities for progress in understanding lie before us. But scientific naturalists claim to know what the form of that progress will be, and to know that mentalistic, teleological, or evaluative intelligibility in particular have been left behind for good as fundamental forms of understanding. It is assumed not only that the natural order is intelligible but that its intelligibility has a certain form, being found in the simplest and most unified physical laws, governing the simplest and fewest elements, from which all else follows. That is what scientific optimists mean by a theory of everything. So long as the basic laws are not themselves necessary truths, the question remains why those laws hold. But perhaps part of the appeal of this conception is that if the laws are simple enough, we can come to rest with them and be content to say that this is just how things are. After all, what is the alternative?

That is really my question. The implausibility of the reductive program that is needed to defend the completeness of this kind of naturalism provides a reason for trying to think of alternatives—alternatives that make mind, meaning, and value as fundamental as matter and space-time in an account of what there is. The fundamental elements and laws of physics and chemistry have been inferred to explain the behavior of the inanimate world. Something more is needed to explain how there can be conscious, thinking creatures whose bodies and brains are composed of those elements. If we want to try to understand the world as a whole, we must start with an adequate range of data, and those data must include the evident facts about ourselves.

4

As a way of marking the boundaries of the territory in which the search for such understanding must proceed, I would now like to say something about the polar opposite of materialism, namely, the position that mind, rather than physical law, provides the fundamental level of explanation of everything, including the explanation of the basic and universal physical laws themselves. This view is familiarly expressed as theism, in its aspect as an explanation of the existence and character of the natural world. It is the most straightforward way of reversing the materialist order of explanation, which explains mind as a consequence of physical law; instead, theism makes physical law a consequence of mind.

Considered as a response to the demand for an all-encompassing form of understanding, theism interprets intelligibility ultimately in terms of intention or purpose—resisting a purely descriptive end point. At the outer bounds of the world, encompassing everything in it, including the law-governed natural order revealed by science, theism places some kind of mind or intention, which is responsible for both the physical and the mental character of the universe. So long as the divine mind just has to be accepted as a stopping point in the pursuit of understanding, it leaves the process incomplete, just as the purely descriptive materialist account does.

For either materialistic or theistic explanation to provide a complete understanding of the world, it would have to be the case that either the laws of physics, or the existence and properties of God and therefore of his creation, cannot conceivably be other than they are. Physicists do not typically believe the former,[2] but theists tend

2. Though Einstein seems to have regarded it as an open question, the question, as he put it, "Did God have any choice when he created the universe?"

to believe the latter. This doesn't mean that a theistic world view must be deterministic: God's essential nature may lead him to create probabilistic laws and beings with free will, whose actions are explained as free choices. But some kind of divine intention would underpin the totality.

The interest of theism even to an atheist is that it tries to explain in another way what does not seem capable of explanation by physical science. The inadequacies of the naturalistic and reductionist world picture seem to me to be real. There are things that science as presently conceived does not help us to understand, and which we can see, from the internal features of physical science, that it is not going to explain. They seem to call for a more uncompromisingly mentalistic or even normative form of understanding. Theism embraces that conclusion by attributing the mental phenomena found within the world to the working of a comprehensive mental source, of which they are miniature versions.

However, I do not find theism any more credible than materialism as a comprehensive world view. My interest is in the territory between them. I believe that these two radically opposed conceptions of ultimate intelligibility cannot exhaust the possibilities. All explanations come to an end somewhere. Both theism and materialism say that at the ultimate level, there is one form of understanding. But would an alternative secular conception be possible that acknowledged mind and all that it implies, not as the expression of divine intention but as a fundamental principle of nature along with physical law? Could it take the form of a unified conception of the natural order, even if it tries to accommodate a richer set of materials than the austere elements of mathematical physics? But let me first say a bit more, for dialectical purposes, about the opposition between theism and materialist naturalism and what is lacking in each of them.

5

The place at which the contrast between forms of intelligibility is most vividly presented is in the understanding of ourselves. This is also the setting for the most heated battles over what physical science can and cannot explain. Both theism and evolutionary naturalism are attempts to understand ourselves from the outside, using very different resources. Theism offers a vicarious understanding, by assigning it to a transcendent mind whose purposes and understanding of the world we cannot ourselves fully share, but which makes it possible to believe that the world is intelligible, even if not to us. The form of this transcendent understanding is conceived by extrapolation from the natural psychological self-understanding we have of our own intentions. Evolutionary naturalism, by contrast, extrapolates to everything, including ourselves, a form of scientific understanding that we have developed in application to certain other parts of the world. But the shared ambition of these two approaches, to encompass ourselves in an understanding that arises from but then transcends our own point of view, is just as important as the difference between them.

What, if anything, justifies this common ambition of transcendence? Isn't it sufficient to try to understand ourselves from within—which is hard enough? Yet the ambition appears to be irresistible—as if we cannot legitimately proceed in life just from the point of view that we naturally occupy in the world, but must encompass ourselves in a larger world view. And to succeed, that larger world view must encompass itself.

Any external understanding, however transcendent, begins from our own point of view (how could it not?) and is usually supposed to be consistent with the main outlines of that point of view even if it also provides a basis for significant criticism and revision as well as extension. With respect to human knowledge, for example, both theism and naturalism try to explain how we can rely on our faculties to understand the world around us. At one extreme there is Descartes' theistic validation of perception and scientific reasoning by the proof that God, who is responsible for our faculties, would not systematically deceive us. At the other extreme there is naturalized epistemology, which argues that perceptual and cognitive faculties evolved by natural selection can be expected to be generally reliable in leading us to true beliefs.

Neither of these proposals provides a defense against radical skepticism—the possibility that our beliefs about the world are systematically false. Such a defense would inevitably be circular, since any confidence we could have in the truth of either a theistic or an evolutionary explanation of our cognitive capacities would have to depend on the exercise of those capacities. For theism, this is the famous Cartesian circle; but there is an analogous naturalistic circle.[3] In addition, evolutionary naturalism offers an explanation of our knowledge that is seriously inadequate, when applied to the knowledge-generating capacities that we take ourselves to have. I will return to this claim below.

But even if these two projects of self-understanding do not refute skepticism, I believe there is a legitimate aim of transcendence that is more modest and perhaps more realistic. We may not be able to rule out the skeptical possibility, and we may not be able to ground our normal capacity for understanding on something in which we can have even greater confidence; but it may still be possible to show how we can reasonably retain our natural confidence in the exercise of the understanding, in spite of the apparent contingencies of our

3. See Barry Stroud, *The Significance of Philosophical Scepticism* (Oxford: Clarendon, 1984), ch. 6, "Naturalized Epistemology."

nature and formation. The hope is not to discover a foundation that makes our knowledge unassailably secure but to find a way of understanding ourselves that is not radically self-undermining, and that does not require us to deny the obvious. The aim would be to offer a plausible picture of how we fit into the world.

6

Even in this more modest enterprise both theism and naturalistic reductionism fall short. Theism does not offer a sufficiently substantial explanation of our capacities, and naturalism does not offer a sufficiently reassuring one.

A theistic account has the advantage over a reductive naturalistic one that it admits the reality of more of what is so evidently the case, and tries to explain it all. But even if theism is filled out with the doctrines of a particular religion (which will not be accessible to evidence and reason alone), it offers a very partial explanation of our place in the world. It amounts to the hypothesis that the highest-order explanation of how things hang together is of a certain type, namely, intentional or purposive, without having anything more to say about how that intention operates except what is found in the results to be explained.

The idea is not empty, because any intentional explanation involves some interpretive assumptions, even about God. An intentional agent must be thought of as having aims that it sees as good, so the aims cannot be arbitrary; a theistic explanation will inevitably bring in some idea of value, and a particular religion can make this much more specific, though it also poses the famous problem of evil. To my mind, apart from the difficulty of believing in God, the disadvantage of theism as an answer to the desire for comprehensive understanding is not that it offers no explanations but that it does not do so in the form of a comprehensive account of the natural order. Theism pushes the quest for intelligibility outside the world. If God exists, he is not part of the natural order but a free agent not governed by natural laws. He may act partly by creating a natural order, but whatever he does directly cannot be part of that order.

A theistic self-understanding, for those who find it compelling to see the world as the expression of divine intention, would leave intact our natural confidence in our cognitive faculties. But it would not be the kind of understanding that explains *how* beings like us fit into the world. The kind of intelligibility that would still be missing is intelligibility of the natural order itself—intelligibility from within. That kind of intelligibility may be compatible with some forms of theism—if God creates a self-contained natural order which he then leaves undisturbed. But it is not compatible with direct theistic explanation of systematic features of the world that would seem otherwise to be brute facts—such as the creation of life from dead matter, or the birth of consciousness, or reason. Such interventionist hypotheses amount to a denial that there is a comprehensive natural order. They are in part motivated by a belief that seems to me correct, namely, that there is little or no possibility that these facts depend on nothing but the laws of physics. But another response to this situation is to think that there may be a completely different type of systematic account of nature, one that makes these neither brute facts that are beyond explanation nor the products of divine intervention. That, at any rate, is my ungrounded intellectual preference.

The problem with naturalistic theories is different: Rather than being reassuring but insufficiently explanatory, materialist theories do try to make the natural order internally intelligible by explaining our place in it without reference to anything outside. But the explanations they propose are not reassuring enough. Evolutionary naturalism provides an account of our capacities that undermines their reliability, and in doing so undermines itself. I will have more to say about these problems of reductionism later; here let me sketch them briefly.

Inevitably, when we construct a naturalistic external self-understanding, we are relying on one part of our "sense-making" capacities to create a system that will make sense of the rest. We rely on evolutionary theory to analyze and evaluate everything from our logical and probabilistic cognition to our moral sense. This reflects the view that empirical science is the one secure, privileged form of understanding and that we can trust other forms only to the extent that they can be validated through a scientific account of how and why they work. That still requires reliance on some of our own faculties. But some faculties are thought to merit more confidence than others, and even if we cannot provide them with a noncircular external justification, we must at least believe that they are not undermined by the external account of their sources and operation that is being proposed. A core of cognitive confidence must remain intact, even if some other faculties are rendered doubtful by their evolutionary pedigree.

Structurally, it is still the Cartesian ideal, but with the leading role played by evolutionary theory instead of by an a priori demonstration of divine benevolence. But I agree with Alvin Plantinga that, unlike divine benevolence, the application of evolutionary theory to the understanding of our own cognitive capacities should undermine, though it need not completely destroy, our confidence in them.[4] Mechanisms of belief formation that have selective advantage in the everyday struggle for existence do not warrant our confidence in the construction of theoretical accounts of the world as a whole. I think the evolutionary hypothesis would

4. Alvin Plantinga, *Warrant and Proper Function* (New York: Oxford University Press, 1993), ch. 12.

imply that though our cognitive capacities *could* be reliable, we do not have the kind of reason to rely on them that we ordinarily take ourselves to have in using them directly—as we do in science. In particular, it does not explain why we are justified in relying on them to correct other cognitive dispositions that lead us astray, though they may be equally natural, and equally susceptible to evolutionary explanation. The evolutionary story leaves the authority of reason in a much weaker position. This is even more clearly true of our moral and other normative capacities—on which we often rely to correct our instincts. I agree with Sharon Street that an evolutionary self-understanding would almost certainly require us to give up moral realism—the natural conviction that our moral judgments are true or false independent of our beliefs.[5] Evolutionary naturalism implies that we shouldn't take any of our convictions seriously, including the scientific world picture on which evolutionary naturalism itself depends.

I will defend these claims in later chapters, but here let me say what would follow if they are correct. The failure of evolutionary naturalism to provide a form of transcendent self-understanding that does not undermine our confidence in our natural faculties should not lead us to abandon the search for transcendent self-understanding. There is no reason to allow our confidence in the objective truth of our moral beliefs, or for that matter our confidence in the objective truth of our mathematical or scientific reasoning, to depend on whether this is consistent with the assumption that those capacities are the product of natural selection. Given how speculative evolutionary explanations of human mental faculties are, they seem too weak a ground for putting into question the most basic forms of thought. Our confidence in the

5. Sharon Street, "A Darwinian Dilemma for Realist Theories of Value," *Philosophical Studies* 127, no. 1 (January 2006): 109–66.

truth of propositions that seem evident on reflection should not be shaken so easily (and, I would add, cannot be shaken on these sorts of grounds without a kind of false consciousness).

It seems reasonable to run the test equally in the opposite direction: namely, to evaluate hypotheses about the universe and how we have come into existence by reference to ordinary judgments in which we have very high confidence. It is reasonable to believe that the truth about what kind of beings we are and how the universe produced us is compatible with that confidence. After all, everything we believe, even the most far-reaching cosmological theories, has to be based ultimately on common sense, and on what is plainly undeniable. The priority given to evolutionary naturalism in the face of its implausible conclusions about other subjects is due, I think, to the secular consensus that this is the only form of external understanding of ourselves that provides an alternative to theism—which is to be rejected as a mere projection of our internal self-conception onto the universe, without evidence.

7

Even if neither evolutionary naturalism nor theism provides the kind of comprehensive self-understanding that we are after, this should not threaten our more direct confidence in the operation of our reason, though its appearance in the world remains a mystery. We can continue to hope for a transcendent self-understanding that is neither theistic nor reductionist. But this also means rejecting a third response to the problem that does not seem to me sustainable, though it has distinguished adherents—namely, to give up the project of external self-understanding altogether and instead to limit ourselves to the sufficiently formidable task of understanding our point of view toward the world from within. Physical science is one aspect of this human point of view, but it can exist side by side with the other aspects, without subsuming them. This pluralistic method is what P. F. Strawson calls "descriptive metaphysics,"[6] and it has much in common with Wittgenstein's antimetaphysical conception of the proper task of philosophy.

But while internal understanding is certainly valuable, and an essential precondition of a more transcendent project, I don't see how we can stop there and not seek an external conception of ourselves as well. To refrain we would have to believe that the quest for a single reality is an illusion, because there are many kinds of truth and many kinds of thought, expressed in many different forms of language, and they cannot be systematically combined through a conception of a single world in which all truth is grounded. That is as radical a claim as any of the alternatives.[7]

The question is there, whether we answer it or not. Even if we conclude that the materialist account of ourselves is incomplete—including its development through evolutionary theory—it remains the case that we are products of the long history of the universe since the big bang, descended from bacteria over billions of years of natural selection. That is part of the true external understanding of ourselves. The question is how we can combine it with the other things we know—including the forms of reason on which that conclusion itself is based—in a world view that does not undermine itself.

6. See P. F. Strawson, *Individuals: An Essay in Descriptive Metaphysics* (London: Methuen, 1959).

7. I am very much in sympathy with the following statement by Jaegwon Kim: "Metaphysics is the domain where different languages, theories, explanations, and conceptual systems come together and have their mutual ontological relationships sorted out and clarified. That there is such a common domain is the assumption of a broad and untendentious realism about our cognitive activities. If you believe that there is no such common domain, well, that's metaphysics, too." *Mind in a Physical World: An Essay on the Mind-Body Problem and Mental Causation* (Cambridge, MA: MIT Press, 1998), 66.

Our own existence presents us with the fact that somehow the world generates conscious beings capable of recognizing reasons for action and belief, distinguishing some necessary truths, and evaluating the evidence for alternative hypotheses about the natural order. We don't know how this happens, but it is hard not to believe that there is some explanation of a systematic kind—an expanded account of the order of the world.

If we find it undeniable, as we should, that our clearest moral and logical reasonings are objectively valid, we are on the first rung of this ladder. It does not commit us to any particular interpretation of the normative, but I believe it demands something more. We cannot maintain the kind of resistance to any further explanation that is sometimes called quietism. The confidence we feel within our own point of view demands completion by a more comprehensive view of our containment in the world.

In the meantime, we go on using perception and reason to construct scientific theories of the natural world even though we do not have a convincing external account of why those faculties exist that is consistent with our confidence in their reliability—neither a naturalistic account nor a Cartesian theistic one. The existence of conscious minds and their access to the evident truths of ethics and mathematics are among the data that a theory of the world and our place in it has yet to explain. They are clearly part of what is the case, just as much as the data about the physical world provided by perception and the conclusions of scientific reasoning about what would best explain those data. We cannot just assume that the latter category of thought has priority over the others, so that what it cannot explain is not real.

Since an adequate form of self-understanding would be an alternative to materialism, it would have to include mentalistic and rational elements of some kind. But my thought is that they could belong to the natural world and need not imply a transcendent individual mind, let alone a perfect being. The inescapable fact that has to be accommodated in any complete conception of the universe is that the appearance of living organisms has eventually given rise to consciousness, perception, desire, action, and the formation of both beliefs and intentions on the basis of reasons. If all this has a natural explanation, the possibilities were inherent in the universe long before there was life, and inherent in early life long before the appearance of animals. A satisfying explanation would show that the realization of these possibilities was not vanishingly improbable but a significant likelihood given the laws of nature and the composition of the universe. It would reveal mind and reason as basic aspects of a nonmaterialistic natural order.

This is not just anthropocentric triumphalism. The entire animal kingdom, the endless generations of insects and spiders in their enormous, extravagant populations, all pose this same question about the order of nature. We have not observed life anywhere but on earth, but no natural fact is cosmologically more significant. However much we come to understand, as we are in the process of doing, the chemical basis of life and of its evolution, the phenomenon still calls for a greatly expanded basis for intelligibility.

To sum up: the respective inadequacies of materialism and theism as transcendent conceptions, and the impossibility of abandoning the search for a transcendent view of our place in the universe, lead to the hope for an expanded but still naturalistic understanding that avoids psychophysical reductionism. The essential character of such an understanding would be to explain the appearance of life, consciousness, reason, and knowledge neither as accidental side effects of the physical laws of nature nor as the result of intentional intervention in nature from without but as an unsurprising if not inevitable consequence of the order that governs the natural world from within. That order would have to include physical law, but if life is not just a physical phenomenon, the origin and evolution of life and mind will not be explainable by physics and chemistry alone. An expanded, but still unified, form of explanation will be needed, and I suspect it will have to include teleological elements.

All that can be done at this stage in the history of science is to argue for recognition of the problem, not to offer solutions. But I want to take up some of the obstacles to reduction, and their consequences, in more detail, beginning with the clearest case.

Chapter 3

Consciousness

1

Consciousness is the most conspicuous obstacle to a comprehensive naturalism that relies only on the resources of physical science. The existence of consciousness seems to imply that the physical description of the universe, in spite of its richness and explanatory power, is only part of the truth, and that the natural order is far less austere than it would be if physics and chemistry accounted for everything. If we take this problem seriously, and follow out its implications, it threatens to unravel the entire naturalistic world picture. Yet it is very difficult to imagine viable alternatives.

Let me begin with a brief history of what has brought us to our present predicament. The modern mind-body problem arose out of the scientific revolution of the seventeenth century, as a direct result of the concept of objective physical reality that drove that revolution. Galileo and Descartes made the crucial conceptual division by proposing that physical science should provide a mathematically precise quantitative description of an external reality extended in space and time, a description limited to spatiotemporal primary qualities such as shape, size, and motion, and to laws governing the relations among them. Subjective appearances, on the other hand—how this physical world appears to human perception—were assigned to the mind, and the secondary qualities like color, sound, and smell were to be analyzed relationally, in terms of the power of physical things, acting on the senses, to produce those appearances in the minds of observers. It was essential to leave out or subtract subjective appearances and the human mind—as well as human intentions and purposes—from the physical world in order to permit this powerful but austere spatiotemporal conception of objective physical reality to develop.

However, the exclusion of everything mental from the scope of modern physical science was bound to be challenged eventually. We humans are parts of the world, and the desire for a unified world picture is irrepressible. It seems natural to pursue that unity by extending the reach of physics and chemistry, in light of their great successes in explaining so much of the natural order. These successes have so far taken the form of reduction followed by reconstruction: discovering the basic elements of which everything is composed and showing how they combine to yield the complexity we observe.

It has become clear that our bodies and central nervous systems are parts of the physical world, composed of the same elements as everything else and completely describable in terms of the modern versions of the primary qualities—more sophisticated but still mathematically and spatiotemporally defined. Molecular biology keeps increasing our knowledge of our own physical composition, operation, and development. Finally, so far as we can tell, our mental lives, including our subjective experiences, and those of other creatures are strongly connected with and probably strictly dependent on physical events in our brains and on the physical interaction of our bodies with the rest of the physical world.

Perhaps it is these developments in neurophysiology and molecular biology that have encouraged the hope of including the mind in a single physical conception of the world; at any rate, the consensus in that direction is recent. Descartes thought it couldn't be done—that mind and matter are both fully real and irreducibly distinct, though they interact. In the dualist view, physical science is defined by the exclusion of the mental from its subject matter. There has always been resistance to dualism, but for several centuries after Descartes, it expressed itself primarily through idealism, the view that mind is the ultimate reality and the physical world is in some way reducible to it. This attempt to overcome the division from the direction of the mental extends from Berkeley—who rejected the primary-secondary quality distinction and held that physical things are ideas in the mind of God—to the logical positivists, who analyzed the physical world as a construction out of sense data. Then, in a rapid historical shift whose causes are somewhat obscure, idealism was largely displaced in later twentieth-century analytic philosophy by attempts at unification in the opposite direction, starting from the physical.

Materialism is the view that only the physical world is irreducibly real, and that a place must be found in it for mind, if there is such a thing. This would continue the onward march of physical science, through molecular biology, to full closure by swallowing up the mind in the objective physical reality from which it was initially excluded. The assumption is that physics is philosophically unproblematic, and the main target of opposition is Descartes' dualist picture of the ghost in the machine. The task is to come up with an alternative, and here begins a series of failures.

One strategy for putting the mental into the physical world picture is conceptual behaviorism, offered as an analysis of the real nature of mental concepts. This was tried in several versions. Mental phenomena were identified variously with behavior, behavioral dispositions, or forms of behavioral organization. In another version, associated with Ryle and inspired by Wittgenstein,[1] mental phenomena were not

1. Gilbert Ryle, *The Concept of Mind* (London: Hutchinson, 1949); Ludwig Wittgenstein, *Philosophical Investigations*, trans. G. E. M. Anscombe (New York: Macmillan, 1953).

identified with anything, either physical or nonphysical; the names of mental states and processes were said not to be referring expressions. Instead, mental concepts were explained in terms of their observable behavioral conditions of application—behavioral criteria or assertability conditions rather than behavioral truth conditions.

All these strategies are essentially verificationist, i.e., they assume that all that needs to be said about the content of a mental statement is what would verify or confirm it, or warrant its assertion, from the point of view of an observer. In one way or another, they reduce mental attributions to the externally observable conditions on the basis of which we attribute mental states to others. If successful, this would obviously place the mind comfortably in the physical world.

It is certainly true that mental phenomena have behavioral manifestations, which supply our main evidence for them in other creatures. Yet all these theories seem insufficient as analyses of the mental because they leave out something essential that lies beyond the externally observable grounds for attributing mental states to others, namely, the aspect of mental phenomena that is evident from the first-person, inner point of view of the conscious subject: for example, the way sugar tastes to you or the way red looks or anger feels, each of which seems to be something more than the behavioral responses and discriminatory capacities that these experiences explain. Behaviorism leaves out the inner mental state itself.

In the 1950s an alternative, nonanalytic route to materialism was proposed, one that in a sense acknowledged that the mental is something inside us, of which outwardly observable behavior is merely a manifestation. This was the psycho-physical identity theory, offered by U. T. Place and J. J. C. Smart[2] not as conceptual

2. U. T. Place, "Is Consciousness a Brain Process?" *British Journal of Psychology* 47 (1956): 44–50; J. J. C. Smart, "Sensations and Brain Processes," *Philosophical Review* 68 (1959): 141–56.

analysis but as a scientific hypothesis. It held that mental events are physical events in the brain: $\Psi = \Phi$ (where Ψ is a mental event like pain or a taste sensation and Φ is the corresponding physical event in the central nervous system). Since this is not a conceptual truth, it cannot be known a priori; it is supposed to be a theoretical identity, like "Water = H_2O," and can be confirmed only by the future development of science.

The trouble is that this nonanalytic identity raises a further question: What is it about Φ that makes it also Ψ? It must be some property conceptually distinct from the physical properties that define Φ. *That is required for the identity to be a scientific and not a conceptual truth.*[3] Clearly materialists won't want to give a dualist answer—i.e., that Φ is Ψ because it has a nonphysical property in addition to its physiological ones (e.g., a nonphysical experiential quality). But they have to give some answer, and it has to be an answer that is consistent with materialism. So defenders of the identity theory, in spite of their wish to avoid relying on conceptual analysis, tended to be pulled back into different kinds of analytical behaviorism, in order to analyze the mental character of brain processes in a way that avoided dualism. What makes the brain process a mental process, they proposed, is not an additional intrinsic property but a relational one—a relation to physical behavior.

A causal element was now added to the analysis: "the inner state that typically causes certain behavior and is caused by certain stimuli." This was prompted by the need to explain the two distinct nonsynonymous references to the same thing that occur in a nonconceptual identity statement. Materialists had to explain how "pain" and "brain state" can refer to the same thing even though their

3. This is Max Black's objection to the identity theory, reported in Smart, "Sensations and Brain Processes," 148n11.

meaning is not the same, and to explain this without appealing to anything nonphysical in accounting for the reference of "pain."[4]

These strategies have taken increasingly sophisticated form, under the headings of causal behaviorism, functionalism, and other theories of how mental concepts could refer to states of the brain in virtue of the causal role of those states in controlling the interaction between the organism and its environment. But all such strategies are unsatisfactory for the same old reason: even with the brain added to the picture, they clearly leave out something essential, without which there would be no mind. And what they leave out is just what was deliberately left out of the physical world by Descartes and Galileo in order to form the modern concept of the physical, namely, subjective appearances.

Another problem was subsequently noticed by Saul Kripke. Identity theorists took as their model for $\Psi = \Phi$ other theoretical identities like "Water = H_2O" or "Heat = Molecular Motion." But Kripke argued that those identities are necessary truths (though not conceptual and not a priori), whereas the Ψ/Φ relation appears to be contingent.[5] This was the basis of Descartes' argument for dualism. Descartes said that since we can clearly conceive of the mind existing without the physical body, and vice versa, they can't be one thing.[6]

Consider "Water = H_2O," a typical scientifically discovered theoretical identity. It means that water is *nothing but* H_2O. You can't have H_2O without water, and you don't need anything more than H_2O for water. It's water even if there's no one around to see, feel, or taste it. We ordinarily identify water by its perceptible qualities, but our perceptual experiences aren't part of the water; they are just effects it has on our

4. See D. M. Armstrong, *A Materialist Theory of the Mind* (London: Routledge & Kegan Paul, 1968).

5. Saul A. Kripke, *Naming and Necessity* (Cambridge, MA: Harvard University Press, 1980).

6. René Descartes, *Meditation* 6.

senses. The intrinsic properties of water, its density, electrical conductivity, index of refraction, liquidity between 0 and 100 degrees centigrade, etc., are all fully explained by H_2O and its properties. The physical properties of H_2O are by themselves sufficient for water.

So if Ψ really is Φ in this sense, and nothing else, then Φ by itself, once its physical properties are understood, should likewise be sufficient for the taste of sugar, the feeling of pain, or whatever it is supposed to be identical with. But it doesn't seem to be. It seems conceivable, for any Φ, that there should be Φ without any experience at all. Experience of taste seems to be something extra, contingently related to the brain state—something *produced* rather than constituted by the brain state. So it cannot be identical to the brain state in the way that water is identical to H_2O.

I have given only a brief sketch of the territory. A voluminous and intricate literature has grown up around these problems, but it serves mainly to confirm how intractable they are. The multiple dead ends in the forward march of materialism suggest that the Ψ/Φ dualism introduced at the birth of modern science may be harder to get out of than many people have imagined. It has even led some philosophers to eliminative materialism—the suggestion that mental events, like ghosts and Santa Claus, don't exist at all.[7] But if we don't regard that as an option and still want to pursue a unified world picture, I believe we will have to leave materialism behind. Conscious subjects and their mental lives are inescapable components of reality not describable by the physical sciences.

I suspect that the appearance of contingency in the relation between mind and brain is probably an illusion, and that it is in fact a necessary but nonconceptual connection, concealed from us by

7. See Paul K. Feyerabend, "Mental Events and the Brain," *Journal of Philosophy* 60 (1963): 295–96.

the inadequacy of our present concepts.[8] Major scientific advances often require the creation of new concepts, postulating unobservable elements of reality that are needed to explain how natural regularities that initially appear accidental are in fact necessary. The evidence for the existence of such things is precisely that if they existed, they would explain what is otherwise incomprehensible.

Certainly the mind-body problem is difficult enough that we should be suspicious of attempts to solve it with the concepts and methods developed to account for very different kinds of things. Instead, we should expect theoretical progress in this area to require a major conceptual revolution at least as radical as relativity theory, the introduction of electromagnetic fields into physics—or the original scientific revolution itself, which, because of its built-in restrictions, can't result in a "theory of everything," but must be seen as a stage on the way to a more general form of understanding. We ourselves are large-scale, complex instances of something both objectively physical from outside and subjectively mental from inside. Perhaps the basis for this identity pervades the world.

2

So far I have argued that the physical sciences will not enable us to understand the irreducibly subjective centers of consciousness that are such a conspicuous part of the world. But the failure of reductionism in the philosophy of mind has implications that extend beyond the mind-body problem. Psychophysical reductionism is an essential component of a broader naturalistic program, which cannot

8. I say more about this in "The Psychophysical Nexus," in *Concealment and Exposure, and Other Essays* (New York: Oxford University Press, 2002), 194–235.

survive without it. This naturalistic program is both metaphysical and scientific. It holds both that everything in the world is physical and that everything that happens in the world has its most basic explanation, whether we can come to know it or not, in physical law, as applied to physical things and events and their constituents.

Many—perhaps most—philosophers of mind are still committed to the reductionist project; they think of the difficulties I have described merely as problems that need to be solved in carrying it out successfully. Whoever shares that point of view can regard the argument that follows as a hypothetical one. Its aim is to show that *if* psychophysical reductionism is ruled out, this infects our entire naturalistic understanding of the universe, not only our understanding of consciousness. Beginning with biology, and seeping down to our conception of the basic constituents of reality, it makes the currently standard materialist form of naturalism untenable, even as an account of the physical world, simply because we are parts of that world. I assume this hypothetical conclusion will be welcome to reductionists, since it shows just how extravagant and costly a position antireductionism in the philosophy of mind is.

Reductionists believe the way has been cleared for the completeness of the materialist conception of the world by some form of functional or causal role analysis of the mental, including all the contents of consciousness. A biological—evolutionary—account of the nature and origin of those behavioral capacities and functions by reference to which consciousness can be analyzed then provides the final link with more basic physical science. All this involves a great deal of speculation and evolutionary guesswork, but the general idea of how consciousness is to be included as part of the physical world is clear enough. It is included in virtue of the existence of physical organisms capable of certain kinds of behavioral interaction with the world, which is in turn explained by genetic variation and natural selection. If there is a problem about how materialism can account for the coming into existence of such organisms, it has nothing in particular to do with consciousness but is just the general problem of whether evolutionary theory really does provide the basis for a reduction of biology to chemistry and physics.

But if the program of analyzing consciousness in terms of behavior and its physical causes is not viable, another problem arises. Even if consciousness is something that cannot be analyzed in terms of the purely physical properties of organisms, its appearance still needs to be explained, as part of the larger project of making sense of the world. Further, any such explanation must account for the fact that the appearance of consciousness on earth and the different forms it takes are closely dependent on the evolutionary development of those physical forms of life that have consciousness. We do not know precisely which forms of life these are, but we can be reasonably sure that they extend far beyond our species. The evolution of life must be at least part of the explanation of the development and forms of consciousness.

The problem, then, is this: What kind of explanation of the development of these organisms, even one that includes evolutionary theory, could account for the appearance of organisms that are not only physically adapted to the environment but also conscious subjects? In brief, I believe it cannot be a purely physical explanation. What has to be explained is not just the lacing of organic life with a tincture of qualia but the coming into existence of subjective individual points of view—a type of existence logically distinct from anything describable by the physical sciences alone. If evolutionary theory is a purely physical theory, then it might in principle provide the framework for a physical explanation of the appearance of behaviorally complex animal organisms with central nervous systems. But subjective consciousness, if it is not reducible to something physical, would not be part of this story; it would be left completely unexplained by physical evolution—even if the physical evolution of such organisms is in fact a causally necessary and sufficient condition for consciousness.

The bare assertion of such a connection is not an acceptable stopping point. It is not an explanation to say just that the physical process of evolution has resulted in creatures with eyes, ears, central nervous systems, and so forth, and that it is simply a brute fact of nature that such creatures *are* conscious in the familiar ways. Merely to identify a cause is not to provide a significant explanation, without some understanding of why the cause produces the effect. The claim I want to defend is that, since the conscious character of these organisms is one of their most important features, the explanation of the coming into existence of such creatures must include an explanation of the appearance of consciousness. That cannot be a separate question. An account of their biological evolution must explain the appearance of conscious organisms as such.

Since a purely materialist explanation cannot do this, the materialist version of evolutionary theory cannot be the whole truth. Organisms such as ourselves do not just *happen* to be conscious; therefore no explanation even of the physical character of those organisms can be adequate which is not also an explanation of their mental character. In other words, materialism is incomplete even as a theory of the physical world, since the physical world includes conscious organisms among its most striking occupants.

This problem depends only on the assumption that even though reductionism is false, mind is a biological phenomenon. So long as the mental is irreducible to the physical, the appearance of conscious physical organisms is left unexplained by a naturalistic account of the familiar type. On a purely materialist understanding of biology, consciousness would have to be regarded as a tremendous and inexplicable extra brute fact about the world. If it is to be explained in any sense naturalistically, through the understanding of organic life, something fundamental must be changed in our conception of the natural order that gave rise to life.

What kind of unified conception of the natural world would allow the explanation of the development of living organisms also to explain the development of consciousness? Antireductionism allows us to pose the question, but to answer it requires something more positive. And it cannot consist (merely!) in a revision of the basic concepts of physics, however radical—as happened with the introduction of electromagnetic fields or relativistic space-time. If we continue to assume that we are parts of the physical world and that the evolutionary process that brought us into existence is part of its history, then something must be added to the physical conception of the natural order that allows us to explain how it can give rise to organisms that are more than physical. The resources of physical science are not adequate for this purpose, because those resources were developed to account for data of a completely different kind.

The appearance of animal consciousness is evidently the result of biological evolution, but this well-supported empirical fact is not yet an explanation—it does not provide understanding, or enable us to see why the result was to be expected or how it came about. In this case, unlike that of the appearance of the physical adaptations characteristic of life, an explanation by natural selection based on physical fitness to survive is not sufficient. Selection for physical reproductive fitness may have resulted in the appearance of organisms that are in fact conscious, and that have the observable variety of different specific kinds of consciousness, but there is no physical explanation of why this is so—nor any other kind of explanation that we know of.

To make facts of this kind intelligible, a postmaterialist theory would have to offer a unified explanation of how the physical and the mental characteristics of organisms developed together, and it would have to do so not just by adding a clause to the effect that the mental comes along with the physical as a bonus. The need for an illuminating explanation of the mental outcome pushes back to impose itself on the understanding of the entire process that led to that outcome.

3

I am putting a great deal of weight on the idea of explanation, and the goal of intelligibility at which it aims—a goal that assumes the fundamental intelligibility of the universe, as discussed in the previous chapter. Not everything has an explanation in this sense. Some things that seem to call for explanation, like the deaths in near succession of several close relatives, may just be coincidences, whose components have unrelated explanations. But systematic features of the natural world are not coincidences, and I do not believe that we can regard them as brute facts not requiring explanation. Regularities, patterns, and functional organization call out for explanation—the more so the more frequent they are. When we become aware of such facts, we conclude that there is something we do not know—something which, if we did know it, would render the facts intelligible. And I take it for granted that knowing the immediate cause of some effect does not always make it intelligible—the causation of consciousness by brain activity being a prime example.

Explanation, unlike causation, is not just of an event, but of an event under a description. An explanation must show why it was likely that an event *of that type* occurred. We may know the causes of the deaths of several members of a family in near succession, but that will not explain why several members of that family died, as such, unless there is some relation among the causes of the individual deaths that makes it antecedently likely that they would strike the group—such as a vendetta or a genetic disease.

Another example: There is a physical explanation of why, when I tap "3," "+," "5," and "=" into my pocket calculator, the figure "8" appears on the display screen. But this causal explanation of the shape on the screen is not an explanation of why the device produced the right answer. To explain the result under that description, we must refer to the algorithm governing the calculator, and the intention of the designer to give it a physical realization.

A naturalistic expansion of evolutionary theory to account for consciousness would not refer to the intentions of a designer. But if it aspires to explain the appearance of consciousness *as such*, it would have to offer some account of why the appearance of conscious organisms, and not merely of behaviorally complex organisms, was likely.

The explanation by standard evolutionary theory of the purely physical characteristics of organisms is hard enough even if one disregards consciousness. As I have said earlier, the physical and functional complexity of the results imposes very demanding conditions on a reductionist historical explanation. The theory of natural selection, if it is to rely only on the operation of physical law, has to postulate that there is a purely physical explanation of why it is not unlikely that accidental mutations in the genetic material have generated the range of variation in viable phenotypes needed to permit natural selection to produce the evolutionary history that has actually occurred on earth over the past three billion years.[9] Like any historical

9. Evolutionary biologists now emphasize that the sources of variation are not "random" in the ordinary sense—that the mutations in the genetic material on which natural selection acts are in various ways restricted and not equally likely, and that this contributes significantly to the process. For example, the recent extraordinary discoveries of evolutionary developmental biology (evo-devo) seem to imply much more system and less chance in the sources of genetic variation than had formerly been supposed. But such facts would also have to be explained ultimately by physical principles in a reductionist theory.

explanation, it will embody a great deal of contingency, so the particular history of life will not be explained by evolutionary theory alone. But the contingencies and their effects have to be consistent with the physical character of the theory. And to complete the link with physics, the explanation has to suppose that there is a nonnegligible probability that some sequence of steps, starting from nonliving matter and depending on purely physical mechanisms, could eventually have resulted in a replicating molecule capable of all this, embodying a precise code billions of characters long, together with the ribosomes that translate that code into proteins.[10] It is not enough to say, "Something had to happen, so why not this?" I find the confidence among the scientific establishment that the whole scenario will yield to a purely chemical explanation hard to understand, except as a manifestation of an axiomatic commitment to reductive materialism.[11]

But to explain consciousness, as well as biological complexity, as a consequence of the natural order adds a whole new dimension of difficulty. I am setting aside outright dualism, which would abandon the hope for an integrated explanation. Indeed, substance dualism would imply that biology has no responsibility at all for the existence of minds.[12] What interests me is the alternative hypothesis

10. At this date, many researchers believe that an "RNA world" preceded the appearance of DNA, but the same questions of probability apply.

11. Compare Richard Lewontin, reviewing Carl Sagan's *The Demon-Haunted World* in the *New York Review of Books*, January 9, 1997: "It is not that the methods and institutions of science somehow compel us to accept a material explanation of the phenomenal world, but, on the contrary, that we are forced by our *a priori* adherence to material causes to create an apparatus of investigation and a set of concepts that produce material explanations, no matter how counter-intuitive, no matter how mystifying to the uninitiated. Moreover, that materialism is absolute, for we cannot allow a Divine Foot in the door."

12. But substance dualism would still leave biology with a huge problem similar to the one we are discussing: namely, why has physical evolution produced organisms of a kind capable of being occupied by and interacting with minds? Presumably the answer would depend on assigning to the mental substance a part in the fitness of the successive organisms. But why minds of the appropriate kinds appear and attach themselves to organisms would remain a complete mystery, from an evolutionary point of view.

that biological evolution is responsible for the existence of conscious mental phenomena, but that since those phenomena are not physically explainable, the usual view of evolution must be revised. It is not just a physical process.

If that is so, how much would have to be added to the physical story to produce a genuine explanation of consciousness—one that made the appearance of consciousness, as such, intelligible, as opposed to merely explaining the appearance of certain physical organisms that, as a matter of fact, are conscious? It is not enough simply to add to the physical account of evolution the further observation that different types of animal organisms, depending on their physical constitution, have different forms of conscious life. That would present the consciousness of animals as a mysterious side effect of the physical history of evolution, which explains only the physical and functional character of organisms.

Elliott Sober once suggested to me in that spirit that consciousness might be like the redness of blood—a side effect of functional biological features that has no function in itself, and no direct explanation by natural selection. In that case consciousness would be like a giant spandrel, in the sense of Gould and Lewontin[13] (and a very lucky one for us). But clearly this bare identification of a cause would not be a satisfactory explanation. Without more, it would explain neither why particular organisms are conscious nor why conscious organisms have come to exist at all.

For a satisfactory explanation of consciousness as such, a general psychophysical theory of consciousness would have to be woven into the evolutionary story, one which makes intelligible both (1) why specific organisms have the conscious life they have, and (2) why

13. Stephen Jay Gould and Richard C. Lewontin, "The Spandrels of San Marco and the Panglossian Paradigm: A Critique of the Adaptationist Programme," *Proceedings of the Royal Society of London B* 205:581–98.

conscious organisms arose in the history of life on earth. At this point such a theory is a complete fantasy, but it is still possible to pose some questions about what it would have to accomplish—in particular about the relation between parts (1) and (2) of the explanatory task.

Suppose there were a general psychophysical theory that, if we could discover it, would allow us to understand, for any type of physical organism, why it did or did not have conscious life, and if it did, why it had the specific type of conscious life that it had. This could be called a nonhistorical theory of consciousness. It would accomplish task (1). But I believe that even if such a powerful nonhistorical theory were conjoined with a purely physical theory of how those organisms arose through evolution, the result would not be an explanation of the appearance of consciousness as such. It would not accomplish task (2); it would still leave the appearance of consciousness as an accidental and therefore unexplained concomitant of something else—the genuinely intelligible physical history.

Let me call a conjunctive explanation one in which A explains B and B has as a consequence C. Sometimes such a conjunction will not amount to an explanation of C as such. Suppose C is "the death of several members of the same family," as discussed above. If A gives the independent cause of each of four deaths, B is the sum of those deaths, and they are in fact members of the same family, then C is a consequence of B but it is not explained, as such, by A. We can explain why four people died who are in fact members of the same family without explaining why four members of the same family died.

Or consider the different conjunctive explanation in the case of the pocket calculator. A is the physical explanation of what happens when I tap in "3 + 5 =," which causes B, the display on the screen of the figure "8." It is a further fact that this figure is the symbol for the number 8, and the figures I tapped in are the symbols for a certain sum, so we have the consequence C that the device produced the right answer for the sum entered. But without more, this is merely an assertion, and not yet an explanation of why the calculator gave that answer, or the right answer. Without the further fact that the calculator was designed to embody an arithmetic algorithm and to display its results in Arabic numerals, the physical explanation alone would leave the arithmetical result completely mysterious. It would give the cause of the figure that appeared on the screen, but would not explain the number as such.

The moral seems to be that a conjunctive explanation, going from A to B and B to C, can explain C only if there is some further, internal relation between the way A explains B and the way B explains C. In the case of the family, this would be satisfied if the same rare hereditary disease killed all four people: each of them developed the disease partly because they were members of the same family. In the case of the calculator, the condition is satisfied since the device has the physical structure and function it has precisely in order to embody the arithmetic algorithm.

It isn't enough that C should be the consequence, even the necessary consequence, of B, which is explained by A. There must be something about A itself that makes C a likely consequence. I believe that if A is the evolutionary history, B is the appearance of certain organisms, and C is their consciousness, this means that some kind of psychophysical theory must apply not only nonhistorically, at the end of the process, but also to the evolutionary process itself. That process would have to be not only the physical history of the appearance and development of physical organisms but also a mental history of the appearance and development of conscious beings. And somehow it would have to be one process, making both aspects of the result intelligible.

If, for example, the explanation of nonreducible conscious life were to preserve the basic structure of evolutionary theory, it would probably involve the following: (1) At least in later stages, consciousness per se plays an essential causal role in the survival and reproduction of organisms. (2) The features of consciousness that play this role are somehow genetically transmitted. (3) The genetic variation among individuals which supplies the candidates for natural selection, at least after a certain point, is simultaneously mental and physical variation. (4) Further, and most significant, it seems unavoidable that these mechanisms should be preceded by others in the earlier stages of evolution that create the conditions for their possibility.

This would mean abandoning the standard assumption that evolution is driven by exclusively physical causes. Indeed, it suggests that the explanation may have to be something more than physical all the way down. The rejection of psychophysical reductionism leaves us with a mystery of the most basic kind about the natural order—a mystery whose avoidance is one of the primary motives of reductionism. It is a double mystery: first, about the relation between the physical and the mental in each individual instance, and second, about how the evolutionary explanation of the development of physical organisms can be transformed into a psychophysical explanation of how consciousness developed.

The existence of consciousness is both one of the most familiar and one of the most astounding things about the world. No conception of the natural order that does not reveal it as something to be expected can aspire even to the outline of completeness. And if physical science, whatever it may have to say about the origin of life, leaves us necessarily in the dark about consciousness, that shows that it cannot provide the basic form of intelligibility for this world. There must be a very different way in which things as they are make sense, and that includes the way the physical world is, since the problem cannot be quarantined in the mind.

4

Given this vacancy in our understanding, what kind of explanation does it make sense to imagine? So far I have considered the possibility of additions or modifications to a standard evolutionary explanation, but now I want to consider a broader range of options. All one can do is to describe abstract possibilities, but to begin with, it is clear that any explanation will have two elements: an ahistorical constitutive account of how certain complex physical systems are also mental, and a historical account of how such systems arose in the universe from its beginnings. Evidently the historical account will depend partly on the correct constitutive account, since the latter describes the outcome that the former has to explain. Let me first discuss the constitutive possibilities.

The constitutive account will be either reductive or emergent. A reductive account will explain the mental character of complex organisms entirely in terms of the properties of their elementary constituents, and if we stay with the assumption that the mental cannot be reduced to the physical, this will mean that the elementary constituents of which we are composed are not merely physical.[14] Since we are composed of the same elements as the rest of the universe, this will have extensive and radical consequences, to which I will return below.

An emergent account, by contrast, will explain the mental character of complex organisms by principles specifically linking mental

14. There is a danger of terminological confusion here. I will use "*reductive*" as the general term for theories that analyze the properties of complex wholes into the properties of their most basic elements. I will continue to use "*reductionist*" for the more specific type of reductive theory that analyzes higher-level phenomena exclusively in terms of physical elements and their physical properties. Psychophysical reductionism is an example. The point to keep in mind is that it is possible for an antireductionist theory to be reductive, provided that the elements to which it reduces higher-level phenomena are not exclusively physical. That is the kind of reductive theory I am talking about here.

states and processes to the complex physical functioning of those organisms—to their central nervous systems in particular, in the case of humans and creatures somewhat like them. The difference from a reductive account is that, while the principles do not reduce the mental to the physical, the connections they specify between the mental and the physical are all higher-order. They concern only complex organisms, and do not require any change in the exclusively physical conception of the elements of which those organisms are composed. An emergent account of the mental is compatible with a physically reductionist account of the biological system in which mind emerges.

To qualify as a genuine explanation of the mental, an emergent account must be in some way systematic. It cannot just say that each mental event or state supervenes on the complex physical state of the organism in which it occurs. That would be the kind of brute fact that does not constitute an explanation but rather calls for explanation. But I think we can imagine a higher-order psychophysical theory that would make the connection cease to seem like a gigantic set of inexplicable correlations and would instead make it begin to seem intelligible. Physiological psychologists are only beginning to uncover the systematic dependence of visual experience on events in the visual cortex, for example, but we can imagine that such explorations will lead to a general theory.

Still, this kind of higher-level theory, however empirically accurate, seems unsatisfactory as a final answer to the constitutive question. If emergence is the whole truth, it implies that mental states are present in the organism as a whole, or in its central nervous system, without any grounding in the elements that constitute the organism, except for the physical character of those elements that permits them to be arranged in the complex form that, according to the higher-level theory, connects the physical with the mental. That such purely physical elements, when combined in a certain way, should necessarily produce a state of the whole that is not constituted out of the properties and relations of the physical parts still seems like magic even if the higher-order psychophysical dependencies are quite systematic.

This dissatisfaction with an explanatory stopping place that relates complex structures to complex structures is what underlies the constant push toward reduction in modern science. It is hard to give up the assumption that whatever is true of the complex must be explained by what is true of the elements. That does not mean that new phenomena cannot emerge at higher levels, but the hope is that they can be analyzed through the character and interactions of their more elementary components. Such harmless emergence is standardly illustrated by the example of liquidity, which depends on the interactions of the molecules that compose the liquid. But the emergence of the mental at certain levels of biological complexity is not like this. According to the emergent position now being considered, consciousness is something completely new.

Because such emergence, even if systematic, remains fundamentally inexplicable, the ideal of intelligibility demands that we take seriously the alternative of a reductive answer to the constitutive question—an answer that accounts for the relation between mind and brain in terms of something more basic about the natural order. If such an account were possible, it would explain the appearance of mental life at complex levels of biological organization by means of a general monism according to which the constituents of the universe have properties that explain not only its physical but its mental character. Tom Sorell states the point clearly:

> Even if the mechanisms that produced biological life, including consciousness, are, at some level, the *same* as those that operate in the evolution of the physical universe, it does not follow that those mechanisms are physical just because physical evolution

> preceded biological evolution. Perhaps some transphysical and transmental concept is required to capture both mechanisms. This conjecture stakes out a territory for something sometimes called "neutral monism" in addition to dualist, materialist, and idealist positions.[15]

Sorell is here using "neutral monism" to designate not just a metaphysical position but a type of systematic explanatory theory distinct from traditional materialism. Considered just metaphysically, as an answer to the mind-body problem, monism holds that certain physical states of the central nervous system are also necessarily states of consciousness—their physical description being only a partial description of them, from the outside, so to speak. Consciousness is in that case not, as in the emergent account, an *effect* of the brain processes that are its physical conditions; rather, those brain processes are *in themselves* more than physical, and the incompleteness of the physical description of the world is exemplified by the incompleteness of their purely physical description.

But since conscious organisms are not composed of a special kind of stuff, but can be constructed, apparently, from any of the matter in the universe, suitably arranged, it follows that this monism will be universal. Everything, living or not, is constituted from elements having a nature that is both physical and nonphysical—that is, capable of combining into mental wholes. So this reductive account can also be described as a form of panpsychism: all the elements of the physical world are also mental.[16] However, the sense in which they are mental is so far exhausted by the claim that they

15. Tom Sorell, *Descartes Reinvented* (Cambridge, UK: Cambridge University Press, 2005), 95.
16. For a thorough and illuminating discussion of this topic see Galen Strawson et al., *Consciousness and Its Place in Nature: Does Physicalism Entail Panpsychism?* (Exeter, UK: Imprint Academic, 2006). The book consists of a target essay by Strawson, seventeen commentaries

are such as to provide a reductive account of how their appropriate combinations necessarily constitute conscious organisms of the kind we are familiar with. Any further consequences of their more-than-physical character at the microlevel remain unspecified by this abstract proposal.

5

Having described the difference between the two types of answers, emergent and reductive, to the constitutive question, let me now turn to the historical question, again on the assumption that psychophysical reductionism is false. The prevailing naturalistic answer to the historical question is the materialist version of evolutionary theory, supplemented by a speculative chemical account of the origin of life. The question is: What alternatives to this picture open up if psychophysical reductionism is rejected?

The historical account of how conscious organisms arose in the universe can take one of three forms: it will be either causal (appealing only to law-governed efficient causation), or teleological, or intentional. (1) A causal historical account will hold that the

---

on it by others, and a long reply by Strawson elaborating his defense of panpsychism. See also my essays "Panpsychism," in *Mortal Questions* (Cambridge, UK: Cambridge University Press, 1979), and "The Psychophysical Nexus"; and the acute and historically informed discussion by Charles Hartshorne, "Physics and Psychics: The Place of Mind in Nature," in *Mind in Nature: Essays on the Interface of Science and Philosophy*, ed. John B. Cobb, Jr., and David Ray Griffin (Washington, DC: University Press of America, 1977), 89–96.

Alfred North Whitehead's philosophy of organism also belongs in this company. Whitehead argued that to identify the abstractions of physics with the whole of reality was to commit the fallacy of misplaced concreteness, and that concrete entities, all the way down to the level of electrons, should all be understood as somehow embodying a standpoint on the world. See Whitehead, *Science and the Modern World: Lowell Lectures, 1925* (New York: Macmillan, 1925).

origin of life and its evolution to the level of conscious organisms has its ultimate explanation in the properties of the elementary constituents of the universe, which are also the constituents of conscious organisms, together with any further properties that may emerge as a result of their combination. (If the constitutive account of consciousness is not emergent but reductive, then the causal historical account will also be fully reductive.) (2) A teleological account will hold that in addition to the laws governing the behavior of the elements in every circumstance, there are also principles of self-organization or of the development of complexity over time that are not explained by those elemental laws. (3) An intentional account will hold that although the natural order provides the constitutive conditions for the possibility of conscious organisms, as it provides the conditions for the possibility of jet aircraft, the realization of this possibility was due to intervention by a being (presumably God) who put the constitutive elements together in the right way—perhaps by assembling the genetic material that would result eventually in the evolution of conscious life. Since either a reductive or an emergent constitutive account could be combined with any of the three types of historical account—causal, teleological, or intentional—there are six options. Let me say something about causal accounts before turning to the other possibilities, which depart much more radically from the usual form of scientific explanation.

A causal historical account could be combined with either an emergent or a reductive constitutive account. In the first alternative, the historical account would be restricted to purely physical explanations of the origin and evolution of life until the point at which organisms reached the kind of complexity that is associated with consciousness. After that, the history would be both a physical and a mental one, and if the emergent mental element played an independent causal role, and was not merely epiphenomenal, the causal process would cease to be strictly reductive. But I am interested in the hypothesis of a physically reductive causal history leading at least to a point at which psychophysical emergence occurs—perhaps in creatures with a nervous system, perhaps sooner. This hypothesis would preserve the standard version of physical evolution without change up to the emergence of consciousness.

Earlier I discussed the question whether a physical account of evolutionary history conjoined with a nonhistorical psychophysical theory could really explain the appearance of consciousness, and I concluded that unless there were some further link between the physical history and the psychophysical theory, this would not render the result intelligible, even if it were causally accurate. It would present consciousness as a mysterious side effect of biological evolution—inevitable, perhaps, but inexplicable as such. To explain consciousness, a physical evolutionary history would have to show why it was likely that organisms *of the kind that have consciousness* would arise.

That would be possible if the psychophysical theory governing the emergence of consciousness revealed it to be inseparable from just the kind of physical organization and functioning of animal life whose development a physical evolutionary history purports to explain through natural selection.[17] That would go a long way toward making evolutionary theory an explanation of why conscious life exists. It would imply that conscious organisms have developed through natural selection precisely in virtue of the kinds of physical characteristics that systematically give rise to consciousness, according to the psychophysical theory of emergence. This, then, is one serious option. It has the disadvantage of postulating the brute fact

17. I am indebted here to Sharon Street.

of emergence, not explainable in terms of anything more basic, and therefore essentially mysterious. And it relies on the large assumption that a reductive physical theory could confer sufficient likelihood on the appearance in geological time of the right kind of physical organisms to trigger that emergence. But it might be regarded as the historical account requiring the smallest alteration to the prevailing physical form of naturalism, while nevertheless acknowledging the irreducibility of the mental to the physical.

However, the other type of causal historical account, based on a reductive rather than an emergent constitutive theory, would in principle explain more. In a different way, it might even be said that the least radical departure from materialist reductionism would be a monistic reductive conception that is both constitutive and historical, as physical theory aims to be with respect to the physical world. The question is whether it makes sense.

A comprehensively reductive conception is favored by the belief that the propensity for the development of organisms with a subjective point of view must have been there from the beginning, just as the propensity for the formation of atoms, molecules, galaxies, and organic compounds must have been there from the beginning, in consequence of the already existing properties of the fundamental particles. If we imagine an explanation taking the form of an enlarged version of the natural order, with complex local phenomena formed by composition from universally available basic elements, it will depend on some kind of monism or panpsychism, rather than laws of psychophysical emergence that come into operation only late in the game.

However, it is not clear that this kind of reductive explanation could really render the result intelligible in the way that particle physics or something comparable ostensibly renders the character and cosmological history of the nonliving material world intelligible. The protopsychic properties of all matter, on such a view, are postulated solely because they are needed to explain the appearance of consciousness at high levels of organic complexity. Apart from that nothing is known about them: they are completely indescribable and have no predictable local effects, in contrast to the physical properties of electrons and protons, which allow them to be detected individually. So we have no idea how such a compositional explanation would work. Without something unimaginably more systematic in the way of a reduction, panpsychism does not provide a new, more basic resting place in the search for intelligibility—a set of basic principles from which more complex results can be seen to follow. It offers only the form of an explanation without any content, and therefore doesn't seem to be much of an advance on the emergent alternative.

Yet the proposal is not empty. In its schematic, pre-Socratic way, this sort of monism attempts to recognize the mental as a physically irreducible part of reality while still clinging to the basic form of understanding that has proved so successful in physical theory. This is not just intellectual imitation; it is encouraged by the close connection between minds and bodies. Organisms are physical complexes whose existence and operation seem to call for reductive explanation, and their existence and operation seem largely or wholly responsible for the existence of consciousness. It therefore seems natural to try to fold the explanation of consciousness into the same reductive structure.

On the other hand, the idea of reducing the mind to elementary mental events or particles seems unnatural in a way that physical atomism doesn't. The space-time framework of the physical world makes the physical part-whole relation immediately graspable, geometrically, but we have no comparably clear idea of a part-whole relation for mental reality—no idea how mental states at the level of organisms could be composed out of the properties of microelements, whether those properties are similar in type to our experiential states or different. Yet a mentalistic reductionism would presumably have to find the protomental parts in a monist counterpart of the physical parts of the organism, and would have to include a theory of how they combine into conscious wholes.

It is even more obscure how properties that would explain how conscious beings are constituted out of universal elements could also help to explain how conscious beings have arisen, historically, in virtue of the laws or principles governing the behavior of those elements. If the theory is to be not only constitutively but historically reductive, then the protomental character of the elements would have to play a part in the explanation of how life began and evolved even before the appearance of animal organisms.

It is already a natural part of the monist conception that the protomental features of the basic constituents are not merely passive but are necessarily also active, since this is needed to explain the inseparability of active and passive in the consciousness of ordinary animals. Just as phenomenology and behavior are internally connected in the mental life of organisms, something analogous must be true at the micro level, if monism is correct. So the protomental will have behavioral implications. Furthermore, if a universal monism is correct, it would mean that these psychophysical connections are unbreakable: one cannot have the mental without the physical aspect, or vice versa.[18]

But this doesn't help us to imagine a monist alternative to the materialist history of the origin and evolution of life, prior to the appearance of conscious organisms. Once conscious organisms appear on the scene, we can see how it would go. For example, a reductive monism would imply that certain structures necessarily have visual experience, in a sense that inextricably combines phenomenology

18. See Nagel, "The Psychophysical Nexus," for more discussion.

and capacities for discrimination in the control of action, and that there are no possible structures capable of the same control without the phenomenology. If such structures appeared on the evolutionary menu, they would presumably enhance the fitness of the resulting organisms. In that way the protomental would play a truly explanatory, and not merely epiphenomenal, role in biological evolution.

But that would not explain why such structures formed in the first place. Even if the possibility of a visual system is somehow already implied by the properties of the basic elements, how can a nonmaterialist monism help to explain its appearance in actuality, over geological time? How could the same active principles that account for action and perception in a fully formed organism also account for the original formation of organisms and the generation of viable mutations over evolutionary history? These questions are analogous to those that can be posed with respect to a purely materialistic reductive evolutionary theory, and they seem just as hard for a nonmaterialist theory.

There will be the same problems about explaining the origin of life and the availability of a sufficient supply of viable mutations for natural selection to work on—sufficient to account for the appearance of (now conscious) life as we know it. The kind of monism or panpsychism that would be needed to provide a nonemergent solution to the constitutive problem will not make these historical questions any easier. Chemistry is assumed to play this double role in the standard materialist explanation of both the living operation and the evolutionary history of physical organisms, including the origin of life. That is already highly speculative, but a hypothetical monism that has expanded to encompass the mind is far more speculative, since it says only that there is more to the basic substance of the world than can be captured by physics and chemistry. The object is to recast the explanation of the evolution of animal organisms so that it explains not only their physical character but also their consciousness and its character and functioning. But even if we conclude that the basis of mind must be present in every part of the universe, that offers no hint of how the monistic properties that underlie consciousness in living organisms lead first to the origin of life and eventually to the appearance of conscious systems on the menu of mutations available for natural selection.

Our beliefs about the properties of the physical elements and their constituents are based on what is needed to account for their contemporary observable behavior and interaction and the results of their combination into molecules and larger structures. The materialist form of naturalism assumes that the history of the universe since the big bang, including the origin and evolution of life, can be explained by those same properties. This is a very large assumption, and an analogous assumption would have to underlie the historical hypothesis of a reductive monism, if it too is based on properties of the elements needed to answer the constitutive question in a way that includes consciousness as a physically irreducible feature of certain organisms. Why should those properties make the appearance of such organisms, starting from inorganic matter, at all likely?

The idea of a reductive answer to both the constitutive and the historical questions remains very dark indeed. It seeks a deeper and more cosmically unified explanation of consciousness than an emergent theory, but at the cost of greater obscurity, and it offers no evident advantage with respect to the historical problem of likelihood.

6

Let me comment more briefly on the intentional and teleological alternatives, whose attractions are enhanced by the difficulties facing a causal account. Either answer to the constitutive question can be combined with an intentional answer to the historical question. Suppose, for example, the constitutive truth is reductive. Then if theistic explanations are possible at all, God might have carried out his purpose of creating conscious beings either by assembling them out of elements with protopsychic properties or by creating a universe with the appropriate highly specific initial conditions to give rise to conscious beings through chemical and then biological evolution, entirely by nonteleological laws of interaction among the elements. Purpose would in that case serve only as the outer frame for a reductive system of efficient causation. For theists, this remains an option.

But if we are trying to imagine a secular theory, according to which the historical development of conscious life is fully explained not by intervention but as part of the natural order, there seem to be only two alternatives: either this development itself depends entirely on efficient causation, operating in its later stages through the mechanisms of biological evolution, or there are natural teleological laws governing the development of organization over time, in addition to laws of the familiar kind governing the behavior of the elements.

This is a throwback to the Aristotelian conception of nature, banished from the scene at the birth of modern science.[19] But I have been persuaded that the idea of teleological laws is coherent, and quite different from the idea of explanation by the intentions of a purposive being who produces the means to his ends by choice. In spite of the exclusion of teleology from contemporary science, it certainly shouldn't be ruled out a priori. Formally, the possibility of principles of change over time tending toward certain types of outcomes is coherent, in a world in which the nonteleological laws are

19. Though of course Aristotle did not have our conception of the world's historical evolution over time.

not fully deterministic.[20] But it is essential, if teleology is to form part of a revised natural order, that its laws should be genuinely universal and not just the description of a single goal-seeking process. Since we are acquainted with only one instance of the appearance and evolution of life, we lack a basis for bringing it under universal teleological laws, unless teleological principles can be found operating consistently at much lower levels. But there would have to be such laws for teleology to genuinely explain anything.

Admittedly, the idea of teleological explanation is often associated with the further idea that the outcomes have value, so that it is not arbitrary that those particular teleological principles hold. That in turn poses the question whether an explanation that appeals to value can be understood apart from the purposes of some being who aims at it. Nonpurposive teleology would either have to be value-free or would have to say that the value of certain outcomes can itself explain why the laws hold.[21] In either case, natural teleology would mean that the universe is rationally governed in more than one way—not only through the universal quantitative laws of physics that underlie efficient causation but also through principles which imply that things happen because they are on a path that leads toward certain outcomes—notably, the existence of living, and ultimately of conscious, organisms.

The teleological option is in many ways obscure. I will have more to say about it later. The reductive causal alternative is equally obscure, but if it made sense, it would have the attraction of greater unity than the teleological, for it would mean not only that the elements of which the natural world consists have properties that result in

20. See John Hawthorne and Daniel Nolan, "What Would Teleological Causation Be?" in *Metaphysical Essays*, ed. John Hawthorne (Oxford: Oxford University Press, 2006), 265–83.

21. John Leslie defends the possibility that value can explain existence. See "The Theory That the World Exists because it Should," *American Philosophical Quarterly* 7 (1970): 286–98, and *Infinite Minds: A Philosophical Cosmology* (Oxford: Clarendon, 2001).

conscious organisms when suitably combined, but that those same properties render it not unlikely that such combinations would actually form by some gradual process in the course of cosmological history, given the time available. The constitutive and the historical questions would then be answered by reference to a common set of principles.

7

So far I have posed the problem by emphasizing the irreducibility of conscious experience to the physical. But I have alluded to the fact that human consciousness is not merely passive but is permeated, both in action and in cognition, with intentionality, the capacity of the mind to represent the world and its own aims. It may be more controversial to claim that intentionality cannot be realized in a purely physical universe than that consciousness cannot be. However, if, as I believe, intentionality, thought, and action resist psychophysical reduction and can exist only in the lives of beings that are also capable of consciousness, then they too form part of what a larger explanation of the mental must account for. This subject will be taken up in the following chapters. I believe that the role of consciousness in the survival of organisms is inseparable from intentionality: inseparable from perception, belief, desire, and action, and finally from reason. The generation of the entire mental structure would have to be explained by basic principles, if it is recognized as part of the natural order.

Philosophy cannot generate such explanations; it can only point out the gaping lack of them, and the obstacles to constructing them out of presently available materials. But in contrast to classical dualism, I suggest that we should not renounce the aim of finding an integrated naturalistic explanation of a new kind. Such a theory cannot be approached directly. It would require many stages, over a long period of time, beginning with greatly expanded empirical information about regularities in the relation between conscious states and brain states in ourselves and closely related organisms. Only later could reductive hypotheses be formulated on this evidential base. But I believe that it makes sense to pursue not only neurophysiological but evolutionary research with a certain utopian long-term goal in mind. We should seek a form of understanding that enables us to see ourselves and other conscious organisms as specific expressions simultaneously of the physical and the mental character of the universe. One might object that life is hard enough to understand considered purely as a physical phenomenon, and that the mind can wait. But adding the requirement that any theory of life also has to explain the development of consciousness may not make the problem worse. Perhaps, on the contrary, the added features of the natural order needed to account for mind will in the end contribute to the explanation of life as well. The more a theory has to explain, the more powerful it has to be.

Chapter 4

Cognition

1

I now want to take up a different type of antireductionist argument and its consequences. Consciousness presents a problem for evolutionary reductionism because of its irreducibly subjective character. This is true even of the most primitive forms of sensory consciousness, such as those presumably found in all animals. The problem that I want to take up now concerns mental functions such as thought, reasoning, and evaluation that are limited to humans, though their beginnings may be found in a few other species. These are the functions that have enabled us to transcend the perspective of the immediate life-world given to us by our senses and instincts, and to explore the larger objective reality of nature and value.

I shall assume that the attribution of knowledge to a computer is a metaphor, and that the higher-level cognitive capacities can be possessed only by a being that also has consciousness (setting aside the question whether their exercise can sometimes be unconscious). That already implies that those capacities cannot be understood through physical science alone, and that their existence cannot be explained by a version of evolutionary theory that is physically reductive. But the problem I now want to discuss goes beyond this. It has to do with the nature of these capacities and the relation they put us in to the world. What we take ourselves to be doing when we think about what is the case or how we should act is something that cannot be reconciled with a reductive naturalism, for reasons distinct from those that entail the irreducibility of consciousness. It is not merely the subjectivity of thought but its capacity to transcend subjectivity and to discover what is objectively the case that presents a problem.

Thought and reasoning are correct or incorrect in virtue of something independent of the thinker's beliefs, and even independent of the community of thinkers to which he belongs. We take ourselves to have the capacity to form true beliefs about the world around us, about the timeless domains of logic and mathematics, and about the right thing to do. We don't take these capacities to be infallible, but we think they are often reliable, in an objective sense, and that they can give us knowledge. The natural internal stance of human life assumes that there is a real world, that many questions, both factual and practical, have correct answers, and that there are norms of thought which, if we follow them, will tend to lead us toward the correct answers to those questions. It assumes that to follow those norms is to respond correctly to values or reasons that we apprehend. Mathematics, science, and ethics are built on such norms.

It is difficult to make sense of all this in traditional naturalistic terms. Unless we are prepared to regard most of it as an illusion, this points to a further expansion of our conception of the natural order to include not only the source of phenomenological consciousness—sensation, perception, and emotion—but also the source of our active capacity to think our way beyond those starting points. The question is how to understand mind in its full sense as a product of nature—or rather, how to understand nature as a system capable of generating mind.

The problem does not arise with respect to the basic forms of perceptual, emotional, and appetitive consciousness that we share with many other animals. Those mental functions do put us into a complex relation with the world around us, but they seem in principle susceptible of an evolutionary explanation provided it is somehow transformed from the materialist version into something capable of explaining the conscious character of these functions. As I indicated in the last chapter, if such experiences can somehow be added to the evolutionary menu, their roles in enabling creatures to navigate in the world, avoid dangers, find nourishment and shelter, and reproduce all make them potentially adaptive and therefore candidates for natural selection. Perception and desire have to meet certain standards of accuracy to enable creatures to survive in the world: they have to enable us to respond similarly to things that are similar and differently to things that are different, to avoid what is harmful, and to pursue what is beneficial. For most creatures, however, objectivity extends no farther than this. Their lives are lived in the world of appearances, and the idea of a more objective reality has no meaning.

But once we come to recognize the distinction between appearance and reality, and the existence of objective factual or practical truth that goes beyond what perception, appetite, and emotion tell us, the ability of creatures like us to arrive at such truth, or even to think about it, requires explanation. An important aspect of this explanation will be that we have acquired language and the possibilities of interpersonal communication, justification, and criticism that language makes possible. But the explanation of our ability to acquire and use language in these ways presents problems of the same order, for language is one of the most important normatively governed faculties. To acquire a language is in part to acquire a system of concepts that enables us to understand reality.

I am going to set aside at this point all the problems mentioned earlier about the probability of the origin of life and the sufficiency of random mutation and natural selection to account for the actual evolutionary history of life on earth. The question I want to raise remains even if those problems can be solved for the evolution of plants and lower animals. I will also suppose for the sake of argument that evolutionary theory can be recast in a way that is consistent with antireductionism, so as to make it capable of explaining the appearance of consciousness. The question I now want to pose is whether our cognitive capacities can be placed in the framework of an evolutionary theory that is in this way no longer exclusively materialist, but that retains the Darwinian structure. It is a hypothetical question, since there may not be such a theory. But I will talk as if there were.

The problem has two aspects. The first concerns the likelihood that the process of natural selection should have generated creatures with the capacity to discover by reason the truth about a reality that extends vastly beyond the initial appearances—as we take ourselves to have done and to continue to do collectively in science, logic, and ethics. Is it credible that selection for fitness in the prehistoric past should have fixed capacities that are effective in theoretical pursuits that were unimaginable at the time? The second problem is the difficulty of understanding naturalistically the faculty of reason that is the essence of these activities. I will begin by considering a possible response to the first problem, before going on to the second, which is particularly intractable.

2

The first problem arises only if one presupposes realism about the subject matter of our thought. We want to know how likely it is, for example, that evolution should have given some human beings the capacity to discover, and other human beings the capacity to understand, the laws of physics and chemistry. If there is no real, judgment-independent physical world, no judgment-independent truths of mathematics, and no judgment-independent truths of ethics and practical reason, then there is no problem of explaining how we are able to learn about them. On an antirealist view, scientific or moral truth depends on our systematic cognitive or conative responses rather than being something independent to which our responses may or may not conform. The "worlds" in question are all just human constructions. In that case an explanation of how those responses—including our scientific theories—were formed will not have to explain their objective correctness in order to be acceptable (although it will have to explain their internal coherence).

Antirealism of this kind is a more serious option for the moral than for the scientific case. One can intelligibly hold that moral realism is implausible because evolutionary theory is the best current explanation of our faculties, and an evolutionary account cannot be given of how we would be able to discover judgment-independent moral truth, if there were such a thing.[1] But it would be awkward to abandon scientific realism for analogous reasons, because one would then have to become an antirealist about evolutionary theory as well. This would mean that evolutionary theory is inconsistent with scientific realism and cannot be understood realistically, which seems an excessively strong result. There would be something strange to the point of incoherence about taking scientific naturalism as the ground for antirealism about natural science.

If we leave the assumption of realism in place, the best hope for a naturalistic response to the first problem would be that evolutionary

1. See Sharon Street, "A Darwinian Dilemma for Realist Theories of Value," *Philosophical Studies* 127 (2006): 109–66. I pursue this topic in the next chapter.

theory, and in particular evolutionary psychology, is in fact capable of giving a credible account of the success of our cognitive capacities. For factual knowledge, this is the aim of naturalized epistemology. The goal would be to explain how innate mental capacities that were selected for their immediate adaptive value are also capable of generating, through extended cultural evolutionary history, true theories about a law-governed natural order that there was no adaptive need to understand earlier. The evolutionary explanation would have to be indirect, since scientific knowledge had no role in the selection of the capacities that generated it.

The just-so story would go roughly like this. Even in the wild, it isn't just perception and operant conditioning that have survival value. The capacity to generalize from experience and to allow those generalizations, or general expectations, to be confirmed or disconfirmed by subsequent experience is also adaptive. So is a basic disposition to maintain logical consistency in belief, by modifying beliefs when inconsistencies arise. A further, very important step would be the capacity to correct individual appearances not only by reference to other conflicting appearances of one's own but also by reference to how things appear to other perceivers. That requires recognition of other minds, an ability with obvious adaptive potential. The reach of these capacities can be greatly extended and deliberately exercised with the help of language, which also permits knowledge to be collectively created, accumulated, and transmitted. With language we can hold in our minds and share with others alternative possibilities, and decide among them on the basis of their consistency or inconsistency with further observations. Complex scientific theories that entail empirical predictions are therefore extensions of the highly adaptive capacity to learn from experience—our own and that of others.

This story depends heavily on the supposition of a biological origin of the capacity for nonperceptual representation through language, resulting in the ability to grasp logically complex abstract structures. It is not easy to say how one might decide whether this could be a manifestation of abilities that have survival value in prehistoric everyday life. In view of the mathematical sophistication of modern physical theories, it seems highly unlikely; but perhaps the claim could be defended.

It is even possible to tell a parallel just-so story about the compatibility between evolutionary theory and moral realism. I am not thinking of the familiar appeal to sociobiology, with its essentially nepotistic interpretation of innate altruistic dispositions. I am not even thinking of the explanation through group selection of dispositions to cooperation in social creatures.[2] Rather, I have in mind the discovery of general principles of value by rational means analogous to those used elsewhere. Starting from an understanding of innate desires and aversions as immediate impressions of value—of what is good or bad for ourselves or our kin—the discovery of a larger, principle-governed normative domain, or domain of practical reason, in which these immediately apparent values are situated, can again proceed through the capacity to generalize and the disposition to avoid inconsistency.

Generalization would lead to the recognition of value in possible future experiences, in the means to them, and in the lives of creatures other than ourselves. These values are not extra properties of goodness and badness, but just truths such as the following: If something I do will cause another creature to suffer, that counts against doing it. I can come to see that this is true by generalizing from the evident disvalue of my own suffering, and once I recognize the more general truth, my motives will be altered. If there are objective general norms of conduct, this kind of thinking would

2. See Elliott Sober and David Sloan Wilson, *Unto Others: The Evolution and Psychology of Unselfish Behavior* (Cambridge, MA: Harvard University Press, 1998).

allow us to discover them even if they are no more innate than the laws of physics. As with science, the process of discovery would be impossible without language, interpersonal communication, and cultural memory. In both cases, although the basic capacities employed are adaptive in their simple form, they would permit us to transcend our starting points to discover large domains of truth quite independently of whether such knowledge enhances fitness.

All this is very far-fetched, but no more so than much evolutionary speculation. It requires that mutations and whatever else may be the sources of genotypic variation should generate not only physical structures but phenomenology, desire and aversion, awareness of other minds, symbolic representations, and logical consistency, all having essential roles in the production of behavior. Provided we can assume some global solution to the mind-body problem that allows all this, the rest of the story suggests that knowledge of objective scientific and moral truth, should there be such things, could result from the exercise of capacities that, in more mundane applications, are at least not inimical to survival. There may not be an insuperable problem of improbability, provided we accept the evolutionary framework itself as probable.

3

However, even if we suppose for the sake of argument that some evolutionary explanation of this kind is true, there is a further problem about thinking of our basic reasoning capacities in this way. It emerges if we contrast the attitude we can reasonably take toward our perceptual and appetitive systems with the attitude we can take toward our reasoning. This will lead to the second problem identified above—the difficulty of understanding reason naturalistically.

If we suppose that there is some way to include consciousness in the evolutionary story, then we can understand our visual system, like the visual systems of other species, to have been shaped by natural selection. The specifics of human vision respond to aspects of the world that have been important in the lives of our ancestors. That allows us to continue to rely on the prima facie evidence of our senses, while recognizing that the evidence will sometimes be misleading, selective, or distorted, and that it bears the marks of our particular biological ancestry.

Something similar is possible in our attitudes toward our intuitive judgments of probability, or toward some of our intuitive value judgments (the desire for revenge, for example). We may come to understand those intuitions as rough but useful unreflective responses shaped by natural selection to a fitness-enhancing form in the circumstances in which our forebears lived and died. At the same time, we can recognize that they may need correction or inhibition. Evolutionary self-awareness of this kind is a common feature of our reflective attitudes toward our natural dispositions of hunger, fear, lust, anger, and so forth.

But whenever we take such a reasonable detached attitude toward our innate dispositions, we are implicitly engaged in a form of thought to which we do not at the same time take that detached attitude. When we rely on systems of measurement to correct perception, or probability calculations to correct intuitive expectations, or moral or prudential reasoning to correct instinctive impulses, we take ourselves to be responding to systematic reasons which in themselves justify our conclusions, and which do not get their authority from their biological origins.[3] They *could* not be backed up

3. See the last chapter of my *The Last Word* (New York: Oxford University Press, 1997); I am here continuing the discussion of questions posed there.

in that way. They don't get their authority from their cultural origins, either; on the contrary, the cultural history that has yielded their development is validated as an instance of progress only by the fact that it has led to these methods for increasing the accuracy of our judgments.

Relying on one's vision and relying on one's reason are similar in one respect: in both cases, the reliance is immediate. When I see a tree, I do not infer its existence from my experience any more than I infer the correctness of a logical inference from the fact that I can't help believing the conclusion. However, there is a crucial difference: in the perceptual case I can recognize that I might be mistaken, but on reflection, even if I think of myself as the product of Darwinian natural selection, I am nevertheless justified in believing the evidence of my senses for the most part, because this is consistent with the hypothesis that an accurate representation of the world around me results from senses shaped by evolution to serve that function. That is not a refutation of radical skepticism, since evolutionary theory, like all of science, depends on the evidence of the senses. But it does provide a coherent picture of my place in the world that is consistent with the general reliability of such evidence.

By contrast, in a case of reasoning, if it is basic enough, the only thing to think is that I have grasped the truth directly. I cannot pull back from a logical inference and reconfirm it with the reflection that the reliability of my logical thought processes is consistent with the hypothesis that evolution has selected them for accuracy. That would drastically weaken the logical claim. Furthermore, in the formulation of that explanation, as in the parallel explanation of the reliability of the senses, logical judgments of consistency and inconsistency have to occur without these qualifications, as direct apprehensions of the truth. It is not possible to think, "Reliance on my reason, including my reliance on *this very judgment*, is reasonable because it is consistent with its having an evolutionary explanation." Therefore any evolutionary account of the place of reason presupposes reason's validity and cannot confirm it without circularity.

Eventually the attempt to understand oneself in evolutionary, naturalistic terms must bottom out in something that is grasped as valid in itself—something without which the evolutionary understanding would not be possible. Thought moves us beyond appearance to something that we cannot regard merely as a biologically based disposition, whose reliability we can determine on other grounds. It is not enough to be able to think that *if* there are logical truths, natural selection might very well have given me the capacity to recognize them. That cannot be my ground for trusting my reason, because even that thought implicitly relies on reason in a prior way.

We can suppose that the capacities which enable us to travel far beyond our innate dispositions in representing and responding to the world have appeared in an ancestor and then been preserved in subsequent generations. The appearance of these capacities has to be integrated with the evolutionary process in that they are at least not inimical to fitness, so that they are not extinguished by natural selection. That much seems plausible. But if I am right to think that we can't regard them merely as further instinctive dispositions, some other explanation is needed of what these capacities are.

Just as consciousness cannot be explained as a mere extension or complication of physical evolution, so reason cannot be explained as a mere extension or complication of consciousness. To explain our rationality will require something in addition to what is needed to explain our consciousness and its evidently adaptive forms, something at a different level. Reason can take us beyond the appearances because it has completely general validity, rather than merely local utility. If we have it, we recognize that it can be neither confirmed nor undermined by a theory of its evolutionary origins, nor by any other external view of itself. We cannot distance ourselves from it. That was Descartes' insight.

If such a thing appeared on the evolutionary menu, it could have proven its adaptive value locally. Then, with the help of cultural deployment and development, it might have risen to its current position of critical authority, correcting and often overruling the older promptings of perception, instinct, and intuition, and not subject to correction by anything else. Its entrenchment and eventual sovereignty over older instincts is comprehensible—but only if we can understand how such a thing can exist at all.

4

This is the second problem: What is the faculty that enables us to escape from the world of appearance presented by our prereflective innate dispositions, into the world of objective reality? And what, besides consciousness, do we have to add to the biological story to make sense of such a faculty?

The distinctive thing about reason is that it connects us with the truth directly. Perception connects us with the truth only indirectly. When I see a tree, I see it because it is there, but not just because it is there. Perception is not a form of insight: I do not grasp the presence of the tree immediately, even though it may seem so prior to reflection. Rather I am aware of it because the tree causes a mental effect in me in virtue of the character of my visual system, which we may suppose has been shaped by natural selection to react in this way to light reflected from physical objects. Having such a system, together with other perceptual and motivational dispositions, enables me to survive in the world. So it is only in a complicated and indirect sense that when I see a tree, I see it because it is there.

But suppose I observe a contradiction among my beliefs and "see" that I must give up at least one of them. (I am driving south in the early morning, and the sun rises on my right.) In that case, I see that the contradictory beliefs cannot all be true, and I see it simply because it is the case. I grasp it directly. It is not adequate to say that, faced with a contradiction, I feel the urgent need to alter my beliefs to escape it, which is explained by the fact that avoiding contradictions, like avoiding snakes and precipices, was fitness-enhancing for my ancestors. That would be an indirect explanation of how the impossibility of the contradiction explains my belief that it cannot be true. But even if some of our ancestors were prey to mere logical phobias and instincts, we have gone beyond that: We reject a contradiction just because we see that it is impossible, and we accept a logical entailment just because we see that it is necessarily true.

In ordinary perception, we are like mechanisms governed by a (roughly) truth-preserving algorithm. But when we reason, we are like a mechanism that can see that the algorithm it follows is truth-preserving. Something has happened that has gotten our minds into immediate contact with the rational order of the world, or at least with the basic elements of that order, which can in turn be used to reach a great deal more. That enables us to possess concepts that display the compatibility or incompatibility of particular beliefs with general hypotheses. We have to start by regarding our prereflective impressions as a partial and perspectival view of the world, but we are then able to use reason and imagination to construct candidates for a larger conception that can contain and account for that part. This applies in the domain of value as well as of fact. The process is highly fallible, but it could not even be attempted without this hard core of self-evidence, on which all less certain reasoning depends. In the criticism and correction of reasoning, the final court of appeal is always reason itself.

What this means is that if we hope to include the human mind in the natural order, we have to explain not only consciousness as it enters into perception, emotion, desire, and aversion but also the conscious control of belief and conduct in response to the awareness of reasons—the avoidance of inconsistency, the subsumption of particular cases under general principles, the confirmation or disconfirmation of general principles by particular observations, and so forth. This is what it is to allow oneself to be guided by the objective truth, rather than just by one's impressions. It is a kind of freedom—the freedom that reflective consciousness gives us from the rule of innate perceptual and motivational dispositions together with conditioning. Rational creatures can step back from these influences and try to make up their own minds. I set aside the question whether this kind of freedom is compatible or incompatible with causal determinism, but it does seem to be something that cannot be given a purely physical analysis and therefore, like the more passive forms of consciousness, cannot be given a purely physical explanation either.

If I decide, when the sun rises on my right, that I must be driving north instead of south, it is because I recognize that my belief that I am driving south is inconsistent with that observation, together with what I know about the direction of rotation of the earth. I abandon the belief because I recognize that it couldn't be true. If I put money into a retirement account because the future income it generates will be more valuable to me then than what I could spend it on now, I act because I see that this makes it a good thing to do. If I oppose the abolition of the inheritance tax, it is because I recognize that the design of property rights should be sensitive not only to autonomy but also to fairness. As the saying goes, I operate in the space of reasons.

The appearance of reason and language in the course of biological history seems, from the point of view of available forms of explanation, something radically emergent—if, as I assume, it cannot be understood behavioristically. Like consciousness, it presents problems of both constitutive and historical explanation. It appeared long after the emergence of conscious creatures, yet it also seems to be essentially a development of consciousness and ought to be understandable as part of that history. Like consciousness, reason is inseparable from the physical life of organisms that have it, since it acts on the material provided by perception and natural desire and controls action, both directly and indirectly. Any understanding of it will transform our understanding of physical organisms and their development as well.

The great cognitive shift is an expansion of consciousness from the perspectival form contained in the lives of particular creatures to an objective, world-encompassing form that exists both individually and intersubjectively. It was originally a biological evolutionary process, and in our species it has become a collective cultural process as well. Each of our lives is a part of the lengthy process of the universe gradually waking up and becoming aware of itself.

5

This, then, is what a theory of everything has to explain: not only the emergence from a lifeless universe of reproducing organisms and their development by evolution to greater and greater functional complexity; not only the consciousness of some of those organisms and its central role in their lives; but also the development of consciousness into an instrument of transcendence that can grasp objective reality and objective value.

Certain things can be assumed, if there is such a thing as reason. First, there are objective, mind-independent truths of different kinds: factual truths about the natural world, including scientific laws; eternal and necessary truths of logic and mathematics; and evaluative and moral truths. Second, by starting from the way things initially appear to us, we can use reason collectively to achieve justified beliefs about some of those objective truths—though some of those beliefs will probably be mistaken. Third, those beliefs in combination can directly influence what we do. Fourth, these processes of discovery and motivation, while mental, are inseparable from physical processes in the organism.

It is trivially true that if there are organisms capable of reason, the possibility of such organisms must have been there from the beginning. But if we believe in a natural order, then something about the world that eventually gave rise to rational beings must explain this possibility. Moreover, to explain not merely the possibility but the actuality of rational beings, the world must have properties that make their appearance not a complete accident: in some way the likelihood must have been latent in the nature of things. So we stand in need of both a constitutive explanation of what rationality might consist in, and a historical explanation of how it arose; and both explanations must be consistent with our being, among other things, physical organisms. The understanding of biological organisms and their evolutionary history would have to expand to accommodate this additional explanatory burden, as I have argued it must expand beyond materialism to accommodate the explanation of consciousness.

Such an explanation would complete the pursuit of intelligibility by showing how the natural order is disposed to generate beings capable of comprehending it. But the obstacles seem enormous. In light of the remarkable character of reason, it is hard to imagine what a naturalistic explanation of it, either constitutive or historical, could look like.

In the previous chapter I explored the possibility of a reductive account of consciousness, based on some form of universal monism or panpsychism. This is modeled on the physical reductionism encouraged by molecular biology, but with an expanded metaphysical basis, in which the physical and the mental are ontologically inseparable. Although it would be a radical departure from the reigning materialist view of nature, the monism required for a reductive but not physically reductionist account of consciousness seems at least conceivable. In answer to the constitutive question, the idea that a complex subject of consciousness might be built up out of minimal protomental elements that are somehow unified simultaneously into an organism and a self has enough potential to merit consideration. Considered as an alternative to an equally speculative emergence of consciousness at high levels of physical organization, it seems relatively credible, in spite of serious problems about the mental part-whole relationship.

However, a reductive account of reason, entirely in terms of the properties of the elementary constituents of which organisms are made, is even more difficult to imagine than a reductive account of consciousness. Rationality, even more than consciousness, seems necessarily a feature of the functioning of the whole conscious subject, and cannot be conceived of, even speculatively, as composed of countless atoms of miniature rationality. The metaphor of the mind as a computer built out of a huge number of transistor-like homunculi will not serve the purpose, because it omits the understanding of the content and grounds of thought and action essential to reason. It could account for behavioral output, but not for understanding. For these reasons, a holistic or emergent answer to the constitutive question comes to seem increasingly more likely than a reductive one as we move up from physical organisms, to consciousness, to reason. This would mean that reason is an irreducible faculty of the kind of fully formed conscious mind that exists in higher animals, and that it cannot be analyzed into the activity of the mind's protomental parts, in the way that sensation perhaps can be.

But the historical question remains. Even if something entirely new begins to happen when the conscious brain reaches a certain size and level of complexity, an explanation of the existence of that complexity will be adequate only if it also explains the existence of reason as such. (This parallels the demands on an explanation of consciousness as such, discussed in the last chapter.) Suppose we have reason because our brains have reached a level of complexity at which reason emerges. If this is to be an explanation that renders the appearance of reason not a complete accident, it must in some way account not just for the physical complexity itself but for the appearance of just the kind of complexity that is a condition of the emergence of reason. This would not be necessary if one were willing to regard reason as a fluke—a pure side effect of other brain developments. But if that is not acceptable, then an explanation of reason would have to explain the likelihood of the appearance of its biological conditions *qua* conditions of reason, i.e., under that description. The possibilities at this point are too abstractly described to permit any speculation as to whether a reductive causal explanation could do this, but if emergence is the correct answer to the constitutive question about reason, it may be that the historical question will require either a teleological or an intentional solution.

6

I have raised the possibility of teleological principles as part of the natural order in the previous chapter. Teleological explanation may have serious problems, but in this case they are no more serious than those of the alternatives, so the possibility should not be disregarded. The evolution of mind is part of a single long process of evolutionary descent. It is the latest stage in the evolution of physical organisms, some of which are now governed largely by thought. If we are skeptical about an intentional (theistic) explanation of the existence of reason, and can't make sense of a causal reductionist one, it is natural to speculate that some tendencies in this direction have been at work all along. If physics alone or even a nonmaterialist monism can't account for the later stages of our evolutionary history, we shouldn't assume that it can account for the earlier stages. Indeed, when we go back far enough, to the origin of life—of self-replicating systems capable of supporting evolution by natural selection—those actually engaged in research in the subject recognize that they are very far from even formulating a viable explanatory hypothesis of the traditional materialist kind. Yet they assume that there must be such an explanation, since life cannot have arisen purely by chance.[4]

In fact, that assumption may be based on a confusion. In an important paper, Roger White has argued that the search for an explanation of the origin of life in terms of the nonpurposive principles of physics and chemistry—an explanation that will reveal that the origin of life is not merely a matter of chance but something to be expected, or at least not surprising—is probably motivated by the sense that life can't be a matter of chance because it looks so much as though it is the product of intentional design. But the hypothesis of intentional design is ruled out as unscientific. So it seems natural to conclude that the only way left for life not to be a matter of chance is for it somehow to be made likely by physical law.

4. See note 8 of chapter 1 for relevant references, and also Iris Fry, *The Emergence of Life on Earth: A Historical and Scientific Overview* (New Brunswick, NJ: Rutgers University Press, 2000).

But as White points out, this inference is illegitimate. Here is what he says:

> The line of reasoning . . . is something like the following. That molecular replicating systems appear to be designed by an agent is sufficient to convince us that they didn't arise by chance. But in scientific reasoning, non-intentional explanations are to be preferred, if possible (some would say at all costs), to intentional ones—hence the motivation to find a non-intentional explanation of life.
>
> It should be clear however, that even granting the appropriateness of a preference for non-intentional explanations, this line of reasoning is confused. In general, if B_I [the hypothesis that the process that led to S was intentionally biased] raises the likelihood of S, then S confirms B_I to at least some degree, and may thereby disconfirm C [the chance hypothesis]. But it does not follow that S confirms B_N [the hypothesis that the process was non-intentionally biased] one iota. S confirms B_N *only if* B_N raises the likelihood of S. If the reason we doubt the Chance Hypothesis is that we suspect that life is due in part to intelligent agency, this by itself gives us no reason to expect there to be a non-intentional explanation for life. If on reflection we do not find the hypothesis of intentional biasing acceptable, then we are left with no reason at all to doubt that life arose by chance.[5]

Much of White's paper is taken up with arguing that life is no more to be expected on the assumption of B_N—the hypothesis of nonintentional bias—than on the assumption of chance. That is because

5. Roger White, "Does Origins of Life Research Rest on a Mistake?" *Noûs* 41 (2003), 475.

even if there is nonintentional bias toward certain outcomes resulting from purposeless physical law, it could be a bias toward any type of outcome whatever, so it cannot make the appearance of life more likely than anything else. As White says,

> What makes certain molecular configurations stand out from the multitude of possibilities seems to be that they are capable of developing into something that strikes us as rather marvelous, namely a world of living creatures. But there is no conceivable reason that blind forces of nature or physical attributes should be biased toward the marvelous.[6]

By contrast, intentional bias is limited as a hypothesis by some rough assumptions about the motives that give rise to intentions. (Thus one cannot claim about just any outcome S, however random or arbitrary, that it gives evidence that the process that led to it was intentionally biased, simply on the ground that it is rendered likely by the hypothesis that it was produced by a being who wanted precisely S to occur. Any argument from design depends on more restrictive general assumptions about what kinds of things a designer might want to produce.)

I am drawn to a fourth alternative, natural teleology, or teleological bias, as an account of the existence of the biological possibilities on which natural selection can operate. I believe that teleology is a naturalistic alternative that is distinct from all three of the other candidate explanations: chance, creationism, and directionless physical law. To avoid the mistake that White finds in the hypothesis of nonintentional bias, teleology would have to be restrictive in what it makes likely, but without depending on intentions or motives. This would probably have to involve some conception of an increase in value through the

6. Ibid., 467.

expanded possibilities provided by the higher forms of organization toward which nature tends: not just any outcome could qualify as a *telos*. That would make value an explanatory end, but not one that is realized through the purposes or intentions of an agent. Teleology means that in addition to physical law of the familiar kind, there are other laws of nature that are "biased toward the marvelous."

The idea of teleology as part of the natural order flies in the teeth of the authoritative form of explanation that has defined science since the revolution of the seventeenth century. Teleology would mean that some natural laws, unlike all the basic scientific laws discovered so far, are temporally historical in their operation. The laws of physics are all equations specifying universal relations that hold at every time and place among mathematically specifiable quantities like force, mass, charge, distance, and velocity. In a nonteleological system the explanation of any temporally extended process has to consist in the explanation, by reference to those laws, of how each state of the universe evolved from its immediate predecessor. Teleology, by contrast, would admit irreducible principles governing temporally extended development.

The teleology I want to consider would be an explanation not only of the appearance of physical organisms but of the development of consciousness and ultimately of reason in those organisms. But its form can be described even if we stay at the physical level. Natural teleology would require two things. First, that the nonteleological and timeless laws of physics—those governing the ultimate elements of the physical universe, whatever they are—are not fully deterministic. Given the physical state of the universe at any moment, the laws of physics would have to leave open a range of alternative successor states, presumably with a probability distribution over them.

Second, among those possible futures there will be some that are more eligible than others as possible steps on the way to the formation of more complex systems, and ultimately of the kinds of replicating systems characteristic of life. The existence of teleology requires that successor states in this subset have a significantly higher probability than is entailed by the laws of physics alone—simply because they are on the path toward a certain outcome. Teleological laws would assign higher probability to steps on paths in state space that have a higher "velocity" toward certain outcomes.[7] They would be laws of the self-organization of matter, essentially—or of whatever is more basic than matter.

This is a frankly teleological hypothesis because the preferred transitions do not have a higher probability in virtue of their intrinsic immediate characteristics, but only in virtue of temporally extended developments of which they form a potential part. In other words, some laws of nature would apply directly to the relation between the present and the future, rather than specifying instantaneous functions that hold at all times. A naturalistic teleology would mean that organizational and developmental principles of this kind are an irreducible part of the natural order, and not the result of intentional or purposive influence by anyone. I am not confident that this Aristotelian idea of teleology without intention makes sense, but I do not at the moment see why it doesn't.

7

What are the alternatives? Any alternative must include the constitutive possibility, in the character of the elements of which the world is composed, of their combination into living organisms with the properties of consciousness, action, and cognition which we know they have. But given this possibility, the historical question of why such

7. See Hawthorne and Nolan, "What Would Teleological Causation Be?"

organisms arose could in principle receive two very different nonteleological answers. First, there is the hypothesis that the initial appearance of a code-governed replicating system that started the evolutionary process was a cosmic accident, and that subsequent accidental mutations provided the set of successive candidates on which natural selection operated to generate the history of life. This hypothesis makes the outcome too accidental to count as a genuine explanation of the existence of conscious, thinking beings as such.

Second, for theists there is the intentional alternative: divine intervention to create life out of the basic material of the world, and perhaps also to guide the process of evolution by natural selection, through the intentional production and preservation of some of the mutations on which natural selection operates along the way.[8] This could be combined with either a reductive or an emergent answer to the constitutive question. A creationist explanation of the existence of life is the biological analogue of dualism in the philosophy of mind. It pushes teleology outside of the natural order, into the intentions of the creator—working with completely directionless materials whose properties nevertheless underlie both the mental and the physical. If God at some point in the past constructed DNA or one of its predecessors out of its elements, that dispenses with the need for any explanation of the capacity of the elements to organize themselves in this apparently purposive way.

That would require only that the existence of DNA be a physical possibility—in chemical space, so to speak. And if we extend the case to consciousness and reason, it would require that conscious and rational subjects supported by brains of the right kind be mental possibilities. But in the creationist picture, the natural order accounts for the physical possibility of DNA in the same way that it accounts for the physical possibility of an airplane or a telephone or a computer.

8. See Alvin Plantinga, *Where the Conflict Really Lies: Science, Religion, and Naturalism* (New York: Oxford University Press, 2011).

Those possibilities are all explained by physics alone: it is only their actualization that involves a designer, and something analogous would be true for animal consciousness—a surprising way in which the protopsychic elements of the world can be combined. So biological and mental organization are no more part of the natural order in the creationist view than airplanes or telephones are. The laws of nature entail their possibility, but they do not explain their actuality.

My preference for an immanent, natural explanation is congruent with my atheism. But even a theist who believes God is ultimately responsible for the appearance of conscious life could maintain that this happens as part of a natural order that is created by God, but that it does not require further divine intervention. A theist not committed to dualism in the philosophy of mind could suppose that the natural possibility of conscious organisms resides already in the character of the elements out of which those organisms are composed, perhaps supplemented by laws of psychophysical emergence. To make the possibility of conscious life a consequence of the natural order created by God while ascribing its actuality to subsequent divine intervention would then seem an arbitrary complication. Some form of teleological naturalism should for these reasons seem no less credible than an interventionist explanation, even to those who believe that God is ultimately responsible for everything.[9]

9. Are there any other alternatives? Well, there is the hypothesis that this universe is not unique, but that all possible universes exist, and we find ourselves, not surprisingly, in one that contains life. But that is a cop-out, which dispenses with the attempt to explain anything. And without the hypothesis of multiple universes, the observation that if life hadn't come into existence we wouldn't be here has no significance. One doesn't show that something doesn't require explanation by pointing out that it is a condition of one's existence. If I ask for an explanation of the fact that the air pressure in the transcontinental jet is close to that at sea level, it is no answer to point out that if it weren't, I'd be dead. Compare Leslie, *Infinite Minds*, 207. See also Roger White, "Fine-Tuning and Multiple Universes," *Noûs* 34 (2000): 260–76; and Ian Hacking, "The Inverse Gambler's Fallacy: The Argument from Design. The Anthropic Principle Applied to Wheeler Universes," *Mind* 96 (1987): 331–40, for further problems with appealing to multiple universes to explain the character of this one.

Chapter 5

Value

1

The idea of teleology implies some kind of value in the result toward which things tend, even if teleology is separated from intention, and the result is not the goal of an agent who aims at it. But let me set these cosmological speculations aside for the moment and take up the more mundane existence and role of value in our lives. Real value—good and bad, right and wrong—is another of those things, like consciousness and cognition, that seem at first sight incompatible with evolutionary naturalism in its familiar materialist form. In the previous chapter I commented briefly on how evolutionary theory might be reconciled with realism about value, following the model of a supposed reconciliation of scientific realism with evolutionary theory. I expressed doubt about the adequacy of such a reconciliation in either case, but I now want to take up the problem of value and reasons for action in more detail.

The problem of the place of value in the natural world includes but goes beyond the problems of the place of consciousness and of cognition in general, because it has to do specifically with the practical domain—the control and assessment of conduct. It is clear that the existence of value and our response to it depend on consciousness and cognition, since so much of what is valuable consists in or involves conscious experience, and the appropriate responses to what is good and bad, right and wrong, depend on the cognitive recognition of the things that give us reasons for and against. Practical reasoning is a cognitive, largely conscious process. I have argued so far that the reality of consciousness and cognition cannot be plausibly reconciled with traditional scientific naturalism, either constitutively or historically. I believe that value presents a further problem for scientific naturalism. Even against the background of a world view in which consciousness and cognition are somehow given a place in the natural order, value is something in addition, and it has consequences that are comparably pervasive.

To explain this claim, it is necessary to explain what is meant by the reality of value—something far less transparent than the reality of consciousness, and also less clear than the reality of cognition or reason in general. One of the difficulties in this area is to describe a form of value realism that is not weighed down by metaphysical baggage but that is still clearly distinguishable from a sophisticated subjectivist conception of value. Value realism is highly controversial because subjectivist accounts of value are not flagrantly implausible in the way that subjectivist accounts of science or materialist accounts of consciousness are. So let me try to explain the kind of value realism that presents the problem, and how it is to be distinguished from sophisticated forms of subjectivism in the Humean tradition, which allow for antirealist versions of moral objectivity, interpersonal justification, and practical reason.

In simple terms, the subjectivist position that I will contrast with realism is that evaluative and moral truth depend on our motivational dispositions and responses, whereas the realist position is that on the contrary, our responses try to reflect the evaluative truth and can be correct or incorrect by reference to it. For the sake of intellectual and discursive economy, I am going to treat subjectivism as a view about truth and truth conditions. But what I say about it, and about the contrast with realism, should be taken to apply also to the equally common expressivist forms of subjectivism, derived from Hume, according to which value judgments express certain kinds of attitudes or feelings of the subject, rather than being true or false in virtue of such attitudes. Such views can usually be put in terms of assertability conditions rather than truth conditions. Subjectivism about values and morality comes in many varieties, and I hope it will be possible to treat the basic issue without distinguishing among them.

The contrast between subjectivism and realism with regard to the dependence of truth on our responses does not apply to the whole of value. In particular, realists can agree with subjectivists that the value of basic experiences of pleasure and pain, for example, is inseparable from our natural responses of attraction and aversion to them. In these cases, for a realist, appearance and reality coincide. To the question, "Would pain be bad even if we didn't mind it?" the answer is "No" (in fact it wouldn't even be pain). It is only when we move to the evaluation of absent experiences—those in the future or those of other subjects—or to judgments about how to deal with possibilities involving multiple experiences, perhaps with conflicting values, or to judgments about the value of things other than experience, that realism and subjectivism will give clearly divergent accounts. The subjectivist position is that the right answer depends on our attitudes and dispositions; the realist position is that our judgments attempt to identify the right answer and to bring our attitudes into accord with it, whether the question is about pain or anything else.[1]

1. I leave aside the large topic of contingent normative truths, which both realists and subjectivists about value can agree are dependent on our collective responses. Anyone who has lived long enough to watch the language drift away from him will know what I mean. "Disinterested" is coming to mean "uninterested," "precipitous" is coming to mean "precipitate," "criteria" is becoming a singular, and "I can't help but think" and "I could care less" are ubiquitous. Reading with daily exasperation canonical publications like the *New York Times*, I have to recognize my unshakeable conviction that these are mistakes as the product of a contingent set of linguistic norms internalized years ago, which I cannot get rid of (and wouldn't even if I could, since they are of course correct).

The most plausible forms of subjectivism rely on some variant of Hume's conception of the passions, including the moral sense. On this type of view, value judgments in general and moral judgments in particular are grounded in aspects of the motivational system more sophisticated and reflective than the basic appetites and instinctive emotions. Prudential judgments are the manifestation of a calm passion of temporally impartial self-interest that generates an equal desire or aversion for future and present benefit or harm to oneself. Moral judgments manifest a sentiment of impartial benevolence, or in more complex cases an attachment to practices or institutions that advance the general welfare, or the good of everyone. The details of the moral sentiment may be complicated, and it may be subject to social modification, but the essential point is that value judgments are nothing more than the expression of such a sentiment: They can be said to be correct or incorrect only by reference to it, together with the non-evaluative facts which are the object of the judgment. What determines the truth of a value judgment is what would result from the calm, reflective, and fully informed operation of these motivational dispositions.

A realist position, by contrast, denies that the truth depends on our dispositions (above the level of immediate feelings like pleasure and pain) and holds that when our value judgments are correct, it is because our dispositions are in accord with the actual structure and weight of values in the case at hand. A judgment that one should not impose serious harm on someone else for the sake of slight benefit to oneself, for example, is based on the recognition that the reason against imposing the harm is much stronger than the reason for pursuing the benefit, and that the fact that the harm would be suffered by someone else is not a reason to disregard it.

2

But what makes it the case that the interests of others provide us with reasons for action, if not that we are disposed on reflection to be motivated by some degree of benevolence? That question is the natural expression of the subjectivist's skepticism about realism. It is also the first step on the path to a fatal misinterpretation of realism—the misinterpretation that turns it into a metaphysical theory. The question implies that something other than value must make value judgments true or false—something in the world in addition to the empirical facts that count for and against action, such as the fact that if I don't step on the brakes I will run over a dog. Subjectivists think this question for the realist is reasonable because their own theory has an answer to it: they find the ground for the truth of value judgments in psychological facts about human motivational dispositions—something more fundamental than values. (For expressivists the question is avoided by denying that value judgments have truth value, but they too find the ground for these judgments in motivational dispositions.)

This kind of explanation of value is precisely what realists deny. More important, in my conception of moral and evaluative realism, realists do not replace the psychological answer to the question posed by subjectivists with a different answer. Realism is not a metaphysical theory of the ground of moral and evaluative truth. It is a metaphysical position only in the negative sense that it denies that all basic truth is either natural or mathematical. It is metaphysical only if the denial of a metaphysical position like naturalism itself counts as a metaphysical position.[2] But value realism does not maintain that value judgments are made true or false by anything else, natural or supernatural.

2. I am indebted for this point to Matthew Silverstein.

Of course natural facts are what make some value judgments true, in the sense that they are the facts that provide reasons for and against action.[3] In that sense the fact that you will run over a dog if you don't step on the brakes makes it the case that you should step on the brakes. But the general moral truth that licenses this inference—namely that it counts in favor of doing something that it will avoid grievous harm to a sentient creature—is not made true by any fact of any other kind. It is nothing but itself.

The dispute between realism and subjectivism is not about the contents of the universe. It is a dispute about the order of normative explanation. Realists believe that moral and other evaluative judgments can often be explained by more general or basic evaluative truths, together with the facts that bring them into play (the fact about the dog). But they do not believe that the evaluative element in such a judgment can be explained by anything else. That there is a reason to do what will avoid grievous harm to a sentient creature is, in a realist view, one of the kinds of things that can be true in itself, and not because something of a different kind is true. In this it resembles physical truths, psychological truths, and arithmetical and geometrical truths.

Instead of explaining the truth or falsity of value judgments in terms of their conformity to our considered motivational dispositions or moral sense, as the subjectivist does, the realist explains our moral sense as a faculty that aims to identify those facts in our circumstances of choice that count for and against certain courses of action, and to discover how they combine to determine what course would be the right one, or what set of alternatives would be permissible or advisable and what others ruled out. The facts that provide these

3. See T. M. Scanlon, *What We Owe to Each Other* (Cambridge, MA: Harvard University Press, 1998), 56–57.

reasons and justifications are just the familiar ones about what would happen if one did this or that, who would benefit, who would be harmed, who has promised what to whom, and so forth.

But although realism does not add anything to the catalogue of entities or properties that a subjectivist believes to exist in the world, it does hold that certain truths that subjectivists think have to be grounded in something else do not have to be so grounded, but are just true in their own right. After all, whatever one's philosophical views, so long as there is such a thing as truth there must be some truths that don't have to be grounded in anything else. Disagreement over which truths these are defines some of the deepest fault lines of philosophy. To philosophers of an idealist persuasion it is self-evident that physical facts can't just be true in themselves, but must be explained in terms of actual or possible experience, just as it is self-evident to those of a materialist persuasion that mental facts can't just be true in themselves, but must be explained in terms of actual or possible behavior, functional organization, or physiology. The issue over moral realism is of the same kind. Someone who finds an unanswerable challenge in the question, "But what *is* a reason?" thereby reveals a restrictive assumption of the largest scope about what kind of basic truth there can be.[4]

In every area of thought we must rely ultimately on our judgments, tested by reflection, subject to correction by the counterarguments of others, modified by the imagination and by comparison with alternatives. Antirealism is always a conjectural possibility: the question can always be posed, whether there is anything more to truth in a certain domain than our tendency to reach certain conclusions in this way, perhaps in convergence with others. Sometimes,

4. Another way of putting the realist's response is to say that the question embodies a category mistake, in Ryle's sense.

as with grammar or etiquette, the answer is no. For that reason the intuitive conviction that a particular domain, like the physical world, or mathematics, or morality, or aesthetics, is one in which our judgments are attempts to respond to a kind of truth that is independent of them may be impossible to establish decisively. Yet it may be very robust all the same, and not unjustified.

To be sure, there are competing subjectivist explanations of the appearance of mind-independence in the truth of moral and other value judgments. One of the things a sophisticated subjectivism allows us to say when we judge that infanticide is wrong is that it would be wrong even if none of us thought so, even though that second judgment too is still ultimately grounded in our responses. However, I find those quasi-realist, expressivist accounts of the ground of objectivity in moral judgments no more plausible than the subjectivist account of simpler value judgments. These epicycles are of the same kind as the original proposal: they deny that value judgments can be true in their own right, and this does not accord with what I believe to be the best overall understanding of our thought about value.

There is no crucial experiment that will establish or refute realism about value. One ground for rejecting it, the type used by Hume, is simply question-begging: if it is supposed that objective moral truths can exist only if they are like other kinds of facts—physical, psychological, or logical—then it is clear that there aren't any. But the failure of this argument doesn't prove that there *are* objective moral truths. Positive support for realism can come only from the fruitfulness of evaluative and moral thought in producing results, including corrections of beliefs formerly widely held and the development of new and improved methods and arguments over time. The realist interpretation of what we are doing in thinking about these things can carry conviction only if it is a better account than the subjectivist or social-constructivist alternative, and that is always going to be a comparative question and a matter of judgment, as it is about any other domain, whether it be mathematics or science or history or aesthetics.

3

All this is a prelude to the larger question: What are the implications for the natural order of different conceptions of value? My insistence that realism about value is not a metaphysical postulation of extra entities or properties might suggest that realism has no implications for the natural order. However, that is not the case. In essence, I agree with Sharon Street's position that moral realism is incompatible with a Darwinian account of the evolutionary influence on our faculties of moral and evaluative judgment.[5] Street holds that a Darwinian account is strongly supported by contemporary science, so she concludes that moral realism is false. I follow the same inference in the opposite direction: since moral realism is true, a Darwinian account of the motives underlying moral judgment must be false, in spite of the scientific consensus in its favor.

But the implications are much larger than that. Since we are evidently the products of evolution, and ultimately of a cosmic process that led to the development first of unicellular organisms and then of conscious agents before eventually producing intelligent beings

5. Throughout the discussion, I will continue to speak in very general terms of Darwinian explanation, even though Darwinian evolutionary theory is immensely complex and the subject of serious internal controversies. I believe that for present purposes the complexities and disagreements within Darwinian theory do not matter. Also, as in the previous chapter, I will set aside all general doubts about the truth of evolutionary theory, and will suppose for the sake of argument that there is a version of it that can accommodate consciousness.

capable of value judgments, the conception of the natural order that made this process possible must be expanded. An adequate conception of the cosmos must contain the resources to account for how it could have given rise to beings capable of thinking successfully about what is good and bad, right and wrong, and discovering moral and evaluative truths that do not depend on their own beliefs. This is analogous to the previously defended implications for the natural order of the existence of consciousness and cognition, but it goes further.

Some comment is called for on the strange, category-jumping nature of this dispute. Street's argument relies on an empirical scientific claim to refute a philosophical position in metaethics. I, even more strangely, am relying on a philosophical claim to refute a scientific theory supported by empirical evidence. But I do not think the movement of thought is inappropriate in either case. Value judgments and moral reasoning are part of human life, and therefore part of the factual evidence about what humans are capable of. The interpretation of faculties such as these is inescapably relevant to the task of discovering the best scientific or cosmological account of what we are and how we came into existence. What counts as a good explanation depends heavily on an understanding of what it is that has to be explained.

Let me begin with an outline of Street's argument that moral realism is incompatible with Darwinism. This is offered in her paper "A Darwinian Dilemma for Realist Theories of Value,"[6] which is quite intricate and addresses more than one form of realism. But the argument is explicitly applied to what she calls *non-naturalist* versions of value realism, according to which evaluative facts or truths are not reducible to any kind of natural facts and do not require any mysterious "extra" properties in the world, but are

6. *Philosophical Studies* 127 (2006): 109–66.

irreducibly normative facts or truths. That is the kind of realism I have described above.

Street points out that if the responses and faculties that generate our value judgments are in significant part the result of natural selection, there is no reason to expect that they would lead us to be able to detect any mind-independent moral or evaluative truth, if there is such a thing. That is because the ability to detect such truth, unlike the ability to detect mind-independent truth about the physical world, would make no contribution to reproductive fitness. It is not at all implausible that the characteristically evaluative motivational dispositions of human beings, and their capacities for practical reasoning and for interpersonal convergence on practices and forms of justification, can be to a significant extent explained, either directly or indirectly, by Darwinian natural selection. But for this explanation it is completely irrelevant whether those faculties enable us to detect mind-independent moral truth, should there be such a thing, or whether they lead us radically astray.

Street observes that the natural Darwinian explanation of the motives and dispositions that form the starting points of our value judgments, and which we can then modify through the process of reflective equilibrium, is that they have contributed to reproductive fitness not only by aiding individual survival but by promoting the nurture and care of children, deterring aggression, and making social cooperation possible. The mind-independent truth of the resulting judgments has no role to play in the Darwinian story: so far as natural selection is concerned, if there were such a thing as mind-independent moral truth, those judgments could be systematically false.

The same cannot be said of our factual judgments. If there is a mind-independent physical world, the systematic inability to detect the basic truth about our surroundings (setting aside more sophisticated scientific truth) would be disastrous for our reproductive fitness. Realism about the physical world is a fundamental aspect of any Darwinian explanation of our perceptual and cognitive faculties, as well as of our motives and capacities for action. But realism about value is irrelevant.

This requires me to revisit something I said in the previous chapter. There I offered a way in which moral realism could be combined with a Darwinian account of our cognitive faculties, including the faculty of practical reason. I said that if we can understand our prereflective impressions of value—instinctive attractions and aversions, inclinations and inhibitions—as appearances of real value, then the cognitive process of discovering a systematic and consistent structure of general reasons and practical and moral principles can be thought of as a way of moving from appearance to reality in the normative domain. To that extent it would be analogous to the Darwinian account of scientific reason, realistically construed. I then went on to make an argument of a different kind against the acceptability of a Darwinian account of reason; but let me set that aside for the moment, because I want to concentrate on a specific problem for value realism that is due to a difference between the prerational data from which factual reasoning and practical reasoning begin. For convenience I will limit the discussion to pleasure and pain, though there are other examples.

The big disanalogy between the two cases from a Darwinian standpoint is between the prereflective appearances of value and the prereflective appearances of the physical world.[7] A realistic conception of the latter, for example of visual perception, is essential to the Darwinian story: vision contributes to reproductive fitness because it enables us to see what is out there, which is necessary for all kinds of successful functioning. By contrast, the real badness of pain and the ability

7. Cf. Street, "A Darwinian Dilemma," 123–25 and n35.

to recognize that badness are completely superfluous in a Darwinian explanation of our aversion to pain. The aversion to pain enhances fitness solely in virtue of the fact that it leads us to avoid the injury associated with pain, not in virtue of the fact that pain is really bad. So far as natural selection is concerned, pain could perfectly well be in itself good, and pleasure in itself bad, or (more likely) both of them might be in themselves valueless—though we are naturally blind to the fact.

A realist, unlike a subjectivist, cannot dismiss these possibilities as meaningless but must instead believe that they are false—just as a realist about the physical world believes that Descartes' evil demon hypothesis about visual perception is not meaningless but false. A Darwinian account of visual perception entails that it gives us information about the external world and that the evil demon hypothesis is false. A Darwinian account of the origin of our basic desires and aversions, by contrast, has no implications as to whether they are generally reliable perceptions of judgment-independent value, or whether indeed there is such a thing.

4

So I am in agreement with Street that, from a Darwinian perspective, the hypothesis of value realism is superfluous—a wheel that spins without being attached to anything. From a Darwinian perspective our impressions of value, if construed realistically, are completely groundless. And if that is true for our most basic responses, it is also true for the entire elaborate structure of value and morality that is built up from them by practical reflection and cultural development—just as scientific realism would be undermined if we abandoned a realistic interpretation of the perceptual experiences on which science is based. Even a system based on the maintenance of coherence or consistency among one's responses does not need the idea of mind-independent truth about value (as opposed to logic), if the responses that provide its original content refer to nothing beyond themselves.

Nevertheless I remain convinced that pain is really bad, and not just something we hate, and that pleasure is really good, and not just something we like. That is just how they glaringly seem to me, however hard I try to imagine the contrary, and I suspect the same is true of most people. That doesn't mean our visceral responses are infallible, any more than our prereflective visual perceptions are. They are merely the starting points for the exploration of a domain that may require extensive practical and moral reasoning to understand. On the Darwinian account, this must be regarded as an illusion—perhaps an illusion of objectivity that is itself the product of natural selection because of its contribution to reproductive fitness. Indeed, the disposition to ascribe an illusory objectivity to plainly contingent, response-dependent norms, of language and custom for example, seems to be typical of humans, and quite useful. However, in my case the scientific credentials of Darwinism, and these other examples, are not enough to dislodge the immediate conviction that objectivity is not an illusion with respect to basic judgments of value.

But what is the realist alternative? Describing it is tricky, since it is obvious that biologically hard-wired pleasure and pain play a vital role in the fitness of conscious creatures even if their objective value doesn't. The realist position must be that these experiences which have desire and aversion as part of their essence also have positive and negative value in themselves, and that this is evident to us on reflection, even though it is not a necessary part of the evolutionary explanation of why they are associated with certain bodily episodes, such as sex, eating, or injury. They are adaptive, but they are something more than that. While they are not the only things that have objective value, these experiences are among the most conspicuous phenomena by which value enters the universe, and the clearest examples through which we become acquainted with real value.

In the realist interpretation, pleasure and pain have a double nature. In virtue of the attraction and aversion that is essential to them, they play a vital role in survival and fitness, and their association with specific biological functions and malfunctions can be explained by natural selection. But for beings like ourselves, capable of practical reason, they are also objects of reflective consciousness, beginning with the judgment that pleasure and pain are good and bad in themselves and leading on, along with other values, to more systematic and elaborate recognition of reasons for action and principles governing their combination and interaction, and ultimately to moral principles.

If the faculties that generate these judgments could also be given a Darwinian explanation, then the realist interpretation would be refuted, for we would then have no reason to regard them as discoveries of what is true independent of our judgments. The objective goodness of pleasure and badness of pain, and the objective truth or falsity of all our more complex value judgments, would be completely irrelevant to the understanding of these faculties or what we are doing when we exercise them. Therefore on the realist view, although the association of pleasure and pain with sex and injury can be explained by natural selection, their objective value, our capacity to recognize it, and all that follows from this cannot be.

Street concludes that realism cannot be right; I conclude that something is missing from Darwinism, and from the standard biological conception of ourselves. Street's negative conclusion comes with a positive alternative, a form of constructivism.[8] Unfortunately

8. See Sharon Street, "Constructivism about Reasons," *Oxford Studies in Metaethics* 3 (2008): 207–46.

my negative conclusion leaves the positive alternative quite mysterious, and unless it is possible to say more, comparison between the competing views of how to respond to the shared hypothetical premise will be difficult.

5

As with the placement of consciousness and cognition in the natural order, the problem of value and the natural order has both a constitutive and a historical aspect. The constitutive question is about our nature: What kind of beings are we, if realism is true and we do indeed recognize and respond to values and practical reasons that are not just the products of our own responses? The historical question is about our origins: What must the universe and the evolutionary process be like to have generated such beings? Both these questions seem to require some alternative to materialist naturalism and its Darwinian application in biology, but what are the possibilities?

The most conspicuous constitutive consequence of realism would be that human beings are able not only to detect but to be motivated by value. In the case of basic experiential values such as the goodness of pleasure and the badness of pain, an instinctive motivation is built into the experience itself: The desire that it continue is part of pleasure and the desire that it stop is part of pain. But when we think of pleasure or pain that is not actually present—either our own experience at other times or the experience of others—this is not the case. Yet we can be motivated by the recognition that pain is bad, and that there is reason to do what will prevent it, whether for ourselves or for others. Such considerations can get us to resist the immediate, built-in motivation of present pleasure or pain, giving it only its objective value. And of course the same is true of much more complex values such as honesty and dishonesty, justice and injustice, loyalty and betrayal.

What is this motivational capacity? As I have said, it should not be construed in terms of an extra metaphysical component of the world, which exercises a causal influence on us. The features of the world that confer value and provide reasons are ordinary facts about the experiences of people, their relations to one another, and the implications for people's and other creatures' lives of different possible courses of action. A reason for action is an ordinary fact, such as the fact that aspirin will cure your headache, and its being a reason is just its counting in favor of your taking aspirin or my giving you some. My capacity to respond to real values is the capacity, for example, to be motivated to give you aspirin for that reason—because I know that it will cure your headache and I recognize that that counts in favor of it, because headaches are bad. We do this kind of thing all the time. What does it mean?

I believe it involves a conscious control of action that cannot be analyzed as physical causation with an epiphenomenal conscious accompaniment, and that it includes some form of free will—though it is, as always, very obscure what sense to give to that notion. I respond consciously to value when I decide to give you aspirin because I learn that you have a headache and I know that aspirin will make it better. Of course I want your headache to go away, but that too is the result of my recognition that headaches are bad. The explanation of my action refers to these facts about headaches and aspirin in their status as reasons—as counting for and against doing certain things. It is through being recognized as reasons by a value-sensitive agent that they affect behavior. It is not so different from the way in which the recognition of reasons in an argument can explain the formation of a factual belief—as discussed in the last chapter.

My convictions about the traditional problem of free will are incompatibilist, but I am not sure that issue has to be settled for present purposes. If there is a way that conscious motivation by reasons for action can be reconciled with causal determinism of action, either physical or psychological, then value realism may be compatible with determinism. For the moment, I wish only to insist that it is not compatible with a Darwinian conception of how the sources of our motives are determined.

If we leave the issue of determinism aside, the distinctive conception of human beings that is implied by value realism is that they can be motivated by their apprehension of values and reasons, whose existence is a basic type of truth, and that the explanation of action by such motives is a basic form of explanation, not reducible to something of another form, either psychological or physical. I give you aspirin because I know it will relieve your headache, thereby manifesting my recognition that this fact counts in favor of the action. And this is not merely a superficial description of something else, which is the real, underlying explanation.

Human action, in other words, is explained not only by physiology, or by desires, but by judgments. We are the subjects of judgment-sensitive attitudes, in Scanlon's phrase,[9] and those judgments have a subject matter beyond themselves. We exist in a world of values and respond to them through normative judgments that guide our actions. This, like our more general cognitive capacities, is a higher development of our nature as conscious creatures. Perhaps it includes the capacity to respond to aesthetic value as well—construed realistically as a judgment-independent domain which our experiences and judgments reveal to us.

9. Scanlon, *What We Owe to Each Other*, 20.

If we ask again in this case about the choice between reductive and emergent explanations, it seems very unlikely that the exercise of consciousness in evaluative judgments and practical reasoning will lend itself to a reductive analysis, even if a reductive monism explains the existence of consciousness itself, in complex organisms. As with cognition in general, the response to value seems only to make sense as a function of the unified subject of consciousness, and not as a combination of the responses of its parts. In that respect it is different from those experiences of pleasure and pain that provide some of the most important raw material of value: they might be explained reductively through some form of psychophysical monism. But practical reasoning and its influence on action involve the unified conscious subject who sees what he should do—and that suggests an emergent answer to the constitutive question. On the other hand, the problem of how to reconcile the unity of the subject, such as it is, with a reductive account of the mental afflicts psychophysical monism at every level. Perhaps at this extravagantly speculative stage the part-whole problem should not rule out reductivism for practical reason any more decisively than it does for taste sensations.

The most important metaphysical aspect of a realist view of practical reason is that consciousness is not epiphenomenal and passive but that it plays an active role in the world. This is part of the ordinary view we have of ourselves, but it adds to the mere psychophysical irreducibility of experiential consciousness, a further respect in which the materialist form of naturalism fails to provide a complete account of the nature of human beings.[10] Whether practical reason is emergent or reducible to activity at the micro level through some form of psychophysical monism, value realism

10. For a persuasive account of the relation between consciousness and free will, see David Hodgson, *Rationality + Consciousness = Free Will* (New York: Oxford University Press, 2011).

requires consciousness to be active and rules out epiphenomenalism in human action. (A subjectivist view of value may also be incompatible with epiphenomenalism, but that is far from clear: subjectivism might be consistent with the possibility that our conscious value judgments are all side effects of the physiological causes of action.)

Setting this issue aside, we come back to the essential contrast between a subjective and a realist conception of value. Subjectivism interprets our value judgments as an outgrowth and elaboration, made possible by our linguistic and rational capacities, of our natural motivational dispositions, nothing more. Realism interprets them as the result of a process of discovery, starting from initial appearances of value that are comparable to perceptual beliefs and moving (we hope) toward a better understanding of how we should live. If realism is true, practical reason in this sense is one of our cognitive faculties.

6

I have not suggested an answer to the constitutive question from the standpoint of realism, only indicated the conditions such an answer would have to meet. Let me now try to say something equally inconclusive about the historical question. It is clear that unlike realism, subjectivism lends itself, at least speculatively, to a Darwinian account of how creatures capable of having values might have arisen. At least this question does not add anything to the difficulties for such an account already presented by the problems of consciousness and of reason in general. Value begins, on this view, from our desires and inclinations, which are natural facts of animal and human psychology, and higher-level value judgments are motivational elaborations from this base, generated by experience, reflection, and culture.

On a realist view, by contrast, the historical question is much more obscure, partly because the result is obscure. We want to know what has to be added to the standard Darwinian picture to account for the appearance through evolution of creatures like us, who can control their actions in response to reasons. While this capacity must be consistent with the influence of natural selection in that it is not inimical to reproductive fitness, and therefore not liable to be extinguished by natural selection, its appearance on the evolutionary menu would have to be explained by something else.

Value realism must make sense of the fact that the biological evolutionary process and the physical and chemical history that preceded it have given rise to conscious creatures, to the real value that fills their lives and experiences, and ultimately to self-conscious beings capable of judgment-sensitive attitudes who can respond to and be rationally motivated by their awareness of those values. The story includes huge quantities of pain as well as pleasure, so it does not lend itself to an optimistic teleological interpretation. Nevertheless, the development of value and moral understanding, like the development of knowledge and reason and the development of consciousness that underlies both of those higher-order functions, forms part of what a general conception of the cosmos must explain. As I have said, the process seems to be one of the universe gradually waking up.

What is the actual history of value in the world, so far as we are aware of it? Nothing in this domain can be regarded as obvious, but in the broadest sense, it seems to coincide with the history of life. First, with the appearance of life even in its earliest forms, there come into existence entities that *have* a good, and for which things can go well or badly. Even a bacterium has a good in this sense, in virtue of its proper functioning, whereas a rock does not. Eventually in the course of evolutionary history there appear conscious beings, whose experiential lives can go well or badly in ways that are familiar to us. Later some descendants of those beings, capable of reflection and self-consciousness, come to recognize what happens to them as good or bad, and to recognize reasons for pursuing or avoiding those things. They learn to think about how these reasons combine to determine what they should do. And finally they develop the collective capacity to think about reasons they may have that do not depend only on what is good or bad for themselves.

This begins with the lives of other beings like themselves, but the question can be extended to good and bad wherever it is found, in the lives of other conscious creatures, and perhaps even in forms of life devoid of conscious experience. Reasons for action apply only to beings with reason, and value can be recognized as the explanation of the reasons that we have, but the concept of value has a much wider range of application than that. Only beings capable of practical reason can recognize value, but once they recognize it, they find it in the lives of creatures without practical reason. In the broadest sense it is probably coextensive with life, though how much of this value we humans have any reason to care about is a question I will leave open. It seems too simple to hold that only the value in conscious lives generates reasons. As Scanlon says, it would be callous and objectionable to cut down a great old tree just for the fun of trying out one's new chain saw.[11]

Two things must be explained to answer the historical question about value and the valuable: first, the appearance of value in the myriad forms it takes in the variety of lives capable of having a good; and second, the appearance of reasons for action and of those beings capable of recognizing them and acting on them. The first would be explained by whatever explains the existence of life, including the place in animal life of consciousness in its many forms. The second requires something more. Does it require more than an explanation of

11. Scanlon, *What We Owe to Each Other*, 68.

the appearance of reason in a general sense? I believe so, because what has to be explained is the appearance of the capacity to recognize and respond to reasons for action, and not just a general cognitive capacity. Practical reason is a development of the motivational system and of the will, not merely a development of the system for forming beliefs.

If value is tied to life, its content will depend on particular forms of life, and the most salient reasons it gives us will depend, even in a realist conception, on our own form of life. This is how a realist account can accommodate one of the things that make subjectivism seem most plausible, namely the fact that what we find self-evidently valuable is overwhelmingly contingent on the biological specifics of our form of life. Human good and bad depend in the first instance on our natural appetites, emotions, capacities, and interpersonal bonds. If we were more like bees or lions, what seems good to us would be very different, a point that Street emphasizes.[12]

It would be a mistake to try to find a common denominator such as pleasure and pain to accommodate in a single realist conception the diverse values that are generated by all the actual, not to mention imaginable, forms of life. Instead, value must be seen as pluralistic: The domain of real value, if there is such a thing, is as rich and complex as the variety of forms of life, or at least of conscious life. Just as most of these lives are only dimly accessible to our understanding from the inside, so the value they generate, positive and negative, is largely beyond our full appreciation. It is also unclear how far the reasons generated by those values can reach—as is true

12. "Imagine, for instance, that we had evolved more along the lines of lions, so that males in relatively frequent circumstances had a strong unreflective evaluative tendency to experience the killing of offspring that were not his own as 'demanded' by the circumstances, and so that females, in turn, experienced no strong unreflective tendency to 'hold it against' a male when he killed her offspring in such circumstances, on the contrary becoming receptive to his advances soon afterwards." Sharon Street, "A Darwinist Dilemma," 120.

even of the values we recognize in forms of human life other than our own. And who knows what unimaginable forms of life and their associated value exist elsewhere in the universe, unrelated to us by common descent? But since value realism can accommodate agent-relative reasons for action, the recognition of what is objectively valuable in the life of one creature does not automatically settle the question of what reasons it implies for the actions of another.

However, our direct access to value comes from human life, the life of one highly specific type of organism in the specific culture it has created. The human world, or any individual human life, is potentially, and often actually, the scene of incredible riches—beauty, love, pleasure, knowledge, and the sheer joy of existing and living in the world. It is also potentially, and often actually, the scene of horrible misery, but on both sides the value, however specific it may be to our form of life, seems inescapably real. Our susceptibility to many of these goods and evils plays a vital role in our survival and reproductive fitness—sexual pleasure, physical pain, the pangs and satisfactions of hunger and thirst—but they are also good and bad in themselves, and we are able to recognize and weigh these values. Initially we recognize them in our own lives, but it cannot stop there.

In looking for a historical explanation, a realist must suppose that the strongly motivating aspects of life and consciousness appear already freighted with value, even though they find their place in the world through their roles in the lives of the organisms that are their subjects. The pleasures of sex, food, and drink are wonderful, in addition to being adaptive. Value enters the world with life, and the capacity to recognize and be influenced by value in its larger extension appears with higher forms of life. Therefore the historical explanation of life must include an explanation of value, just as it must include an explanation of consciousness.

If we recall the three potential types of historical explanation—causal, teleological, and intentional—it is hard to see how a causal explanation would be possible. Even if there were a partly reductive answer to the constitutive question about the existence of value—if the value of an experience of pleasure were constituted, for example, by the combined value of it protomental parts—that doesn't lead anywhere with regard to the historical question. It is difficult to imagine what form of psychophysical monism could make possible a reductive historical explanation of the origin of life, the development of conscious life, and the appearance of practical reason that would make it anything other than a complete accident that what we care about has objective value.

By contrast, once we recognize that an explanation of the appearance and development of life must at the same time be an explanation of the appearance and development of value, a teleological explanation comes to seem more eligible. This would mean that what explains the appearance of life is in part the fact that life is a necessary condition of the instantiation of value, and ultimately of its recognition.

I will again set aside the hypothesis of an intentional explanation, even though it, too, could meet this condition. That leaves teleology. According to the hypothesis of natural teleology, the natural world would have a propensity to give rise to beings of the kind that have a good—beings for which things can be good or bad.[13] These are all the actual and possible forms of life. They have appeared through the historical process of evolution, but part of the explanation for the existence of that process and of the possibilities on which natural selection operates would be that they bring value into the world, in a great variety of forms.

13. See the description of the logic of such explanation in the previous chapter.

Since the emergence of value is the emergence of both good and evil, it is not a candidate for a purely benign teleological explanation: a tendency toward the good. In fact, no teleological principle tending toward the production of a single outcome seems suitable. Rather, it would have to be a tendency toward the proliferation of complex forms and the generation of multiple variations in the range of possible complex systems.[14]

If we were not inclined to recognize objective reasons for action, and were motivated exclusively by our desires, we would have no reason to believe in the existence of value in a realist sense. There would be nothing to explain, beyond the system of subjective motives and their capacity to be guided by the information delivered by perception, memory, and theoretical reason. But if we take our impressions of objective value to be substantially correct, rather than completely illusory, then we must regard the appearance and evolution of life as something more than a history of the development of self-reproducing organisms, as it is in the Darwinian version.

We recognize that evolution has given rise to multiple organisms that *have* a good, so that things can go well or badly for them, and that in some of those organisms there has appeared the additional capacity to aim consciously at their own good, and ultimately at what is good in itself. From a realist perspective this cannot be merely an accidental side effect of natural selection, and a teleological explanation satisfies this condition. On a teleological account, the existence of value is not an accident, because that is part of the

14. Such a possibility is described by C. D. Broad in *The Mind and Its Place in Nature* (London: Routledge & Kegan Paul, 1925), 81–94. Henri Bergson's conception of creative evolution postulated a similar tendency, but he thought of it not as an addition to the natural order but as the free creation of a universal vital force. See *L'évolution créatrice* (1907); available in English as *Creative Evolution*, trans. Arthur Mitchell (New York: Henry Holt, 1911).

explanation of why there is such a thing as life, with all its possibilities of development and variation. In brief, value is not just an accidental side effect of life; rather, there is life because life is a necessary condition of value.[15]

This is a revision of the Darwinian picture rather than an outright denial of it. A teleological hypothesis will acknowledge that the details of that historical development are explained largely through natural selection among the available possibilities on the basis of reproductive fitness in changing environments. But even though natural selection partly determines the details of the forms of life and consciousness that exist, and the relations among them, the existence of the genetic material and the possible forms it makes available for selection have to be explained in some other way. The teleological hypothesis is that these things may be determined not merely by value-free chemistry and physics but also by something else, namely a cosmic predisposition to the formation of life, consciousness, and the value that is inseparable from them.

In the present intellectual climate such a possibility is unlikely to be taken seriously, but I would repeat my earlier observation that no viable account, even a purely speculative one, seems to be available of how a system as staggeringly functionally complex and information-rich as a self-reproducing cell, controlled by DNA, RNA, or some predecessor, could have arisen by chemical evolution alone from a dead environment. Recognition of the problem is not limited to the defenders of intelligent design. Although scientists continue to seek a purely chemical explanation of the origin of life, there are also

15. Compare Derek Parfit on why the universe exists. "Why Anything? Why This?" *London Review of Books*, January 22, 1998; reprinted in Parfit, *On What Matters* (Oxford: Oxford University Press, 2011).

card-carrying scientific naturalists like Francis Crick who say that it seems almost a miracle.[16] Crick is led by his reflection on the probabilities to the hypothesis of "directed panspermia"—that Earth was seeded with unicellular life sent from an advanced civilization elsewhere in our galaxy where life had evolved earlier. This depends on the supposition that there were other planets of other stars whose physical environment made the accidental formation of life less unlikely. But Crick acknowledges that there is no basis for confidence about any of these likelihoods.

Some form of natural teleology, a type of explanation whose intelligibility I briefly defended in the last chapter, would be an alternative to a miracle—either in the sense of a wildly improbable fluke or in the sense of a divine intervention in the natural order. The tendency for life to form may be a basic feature of the natural order, not explained by the nonteleological laws of physics and chemistry. This seems like an admissible conjecture given the available evidence. And once there are beings who can respond to value, the rather different teleology of intentional action becomes part of the historical picture, resulting in the creation of new value. The universe has become not only conscious and aware of itself but capable in some respects of choosing its path into the future—though all three, the consciousness, the knowledge, and the choice, are dispersed over a vast crowd of beings, acting both individually and collectively.

These teleological speculations are offered merely as possibilities, without positive conviction. What I am convinced of is the negative claim that, in order to understand our questions and judgments

16. Francis Crick, *Life Itself: Its Origin and Nature* (New York: Simon & Schuster, 1981), 88. See also Jacques Monod, *Chance and Necessity: An Essay on the Natural Philosophy of Modern Biology*, trans. Austin Wainhouse (New York: Knopf, 1971); and Ernst Mayr, *Evolution and the Diversity of Life: Selected Essays* (Cambridge, MA: Harvard University Press, 1976).

about values and reasons realistically, we must reject the idea that they result from the operation of faculties that have been formed from scratch by chance plus natural selection, or that are incidental side effects of natural selection, or are products of genetic drift. When we ask ourselves, for example, whether revenge is a true justification or just a natural motive, or what kind of weight we should give to the interests of strangers or of other species, we should think of ourselves as calling on a capacity of judgment that allows us to transcend the imperatives of biology.

I believe this is also true of our use of theoretical reason to determine the real character of the world presented to us by sense perception.[17] As I have said, the judgment that our senses are reliable because their reliability contributes to fitness is legitimate, but the judgment that our reason is reliable because its reliability contributes to fitness is incoherent. That judgment cannot itself depend on this kind of empirical confirmation without generating a regress: to make the judgment is necessarily to take it as having authority in its own right. I don't think any other mental stance is available in the theoretical case.

But in the case of value and practical reason, I believe it is coherent to be subjectivist—to regard all impressions of objective value or objective reasons for action as illusory, and to think of the processes of practical deliberation and moral reasoning as nothing but sophisticated ways of deciding what one really wants. The theoretical use of reason, standing on its own, provides a

17. Sharon Street, interestingly enough, holds that theoretical reason too cannot be interpreted realistically, if our reasoning capacities have to a significant extent an evolutionary explanation. She is a realist about *truth* with respect to the natural world, but not a realist about epistemic reasons. In the domain of value, of course, she is an antirealist about both truth and reasons. See Sharon Street, "Evolution and the Normativity of Epistemic Reasons," *Canadian Journal of Philosophy*, Supplementary Volume 35 (2011): 213–48.

frame within which it is in principle possible to think of one's motivational system as a product of natural causes alone, and never as a response to what is really good and bad. Although I find it impossible to take up this position, I do not think it is unintelligible. The question is one of relative plausibility. And I have to acknowledge that for someone not disposed to accept value realism, the radical consequences I have drawn from it will only increase its implausibility.

Chapter 6

Conclusion

Philosophy has to proceed comparatively. The best we can do is to develop the rival alternative conceptions in each important domain as fully and carefully as possible, depending on our antecedent sympathies, and see how they measure up. That is a more credible form of progress than decisive proof or refutation.

In the present climate of a dominant scientific naturalism, heavily dependent on speculative Darwinian explanations of practically everything, and armed to the teeth against attacks from religion, I have thought it useful to speculate about possible alternatives. Above all, I would like to extend the boundaries of what is not regarded as unthinkable, in light of how little we really understand about the world. It would be an advance if the secular theoretical establishment, and the contemporary enlightened culture which it dominates, could wean itself of the materialism and Darwinism of the gaps—to adapt one of its own pejorative tags. I have tried to show that this approach is incapable of providing an adequate account, either constitutive or historical, of our universe.

However, I am certain that my own attempt to explore alternatives is far too unimaginative. An understanding of the universe as basically prone to generate life and mind will probably require a much more radical departure from the familiar forms of naturalistic explanation than I am at present able to conceive. Specifically, in attempting to understand consciousness as a biological phenomenon, it is too easy to forget how radical is the difference between the subjective and the objective, and to fall into the error of thinking about the mental in terms taken from our ideas of physical events and processes. Wittgenstein was sensitive to this error, though his way of avoiding it through an exploration of the grammar of mental language seems to me plainly insufficient.

It is perfectly possible that the truth is beyond our reach, in virtue of our intrinsic cognitive limitations, and not merely beyond our grasp in humanity's present stage of intellectual development. But I believe that we cannot know this, and that it makes sense to go on seeking a systematic understanding of how we and other living things fit into the world. In this process, the ability to generate and reject false hypotheses plays an essential role. I have argued patiently against the prevailing form of naturalism, a reductive materialism that purports to capture life and mind through its neo-Darwinian extension. But to go back to my introductory remarks, I find this view antecedently unbelievable—a heroic triumph of ideological theory over common sense. The empirical evidence can be interpreted to accommodate different comprehensive theories, but in this case the cost in conceptual and probabilistic contortions is prohibitive. I would be willing to bet that the present right-thinking consensus will come to seem laughable in a generation or two—though of course it may be replaced by a new consensus that is just as invalid. The human will to believe is inexhaustible.

INDEX